T0142472

Springer Theses

Recognizing Outstanding Ph.D. Research

Aims and Scope

The series "Springer Theses" brings together a selection of the very best Ph.D. theses from around the world and across the physical sciences. Nominated and endorsed by two recognized specialists, each published volume has been selected for its scientific excellence and the high impact of its contents for the pertinent field of research. For greater accessibility to non-specialists, the published versions include an extended introduction, as well as a foreword by the student's supervisor explaining the special relevance of the work for the field. As a whole, the series will provide a valuable resource both for newcomers to the research fields described, and for other scientists seeking detailed background information on special questions. Finally, it provides an accredited documentation of the valuable contributions made by today's younger generation of scientists.

Theses are accepted into the series by invited nomination only and must fulfill all of the following criteria

- They must be written in good English.
- The topic should fall within the confines of Chemistry, Physics, Earth Sciences, Engineering and related interdisciplinary fields such as Materials, Nanoscience, Chemical Engineering, Complex Systems and Biophysics.
- The work reported in the thesis must represent a significant scientific advance.
- If the thesis includes previously published material, permission to reproduce this must be gained from the respective copyright holder.
- They must have been examined and passed during the 12 months prior to nomination.
- Each thesis should include a foreword by the supervisor outlining the significance of its content.
- The theses should have a clearly defined structure including an introduction accessible to scientists not expert in that particular field.

More information about this series at http://www.springer.com/series/8790

Jeffrey Roskes

A Boson Learned from its Context, and a Boson Learned from its End

A dissertation submitted to Johns Hopkins University in conformity with the requirements for the degree of Doctor of Philosophy

Jeffrey Roskes
Bloomberg Center for Physics
& Astronomy
Johns Hopkins University
Baltimore, MD, USA

ISSN 2190-5053 ISSN 2190-5061 (electronic)
Springer Theses
ISBN 978-3-030-58013-1 ISBN 978-3-030-58011-7 (eBook)
https://doi.org/10.1007/978-3-030-58011-7

This Springer imprint is published by the registered company Springer Nature Switzerland AG.
The registered company address is: Gewerbestrasse 11, 6330 Cham, Switzerland

This thesis is dedicated to my grandfather, Baba ז"ל. I know you would have loved to see it.

Supervisor's Foreword

The quantum theory of the matter and energy that surrounds us, the so-called standard model (SM) of particle physics, has been extremely successful in describing both the microscopic world around us and some of the earliest moments of our Universe. One of the last breakthroughs in completing the full picture of the SM was the discovery of the Higgs boson at the Large Hadron Collider (LHC) in 2012. Since then, we have been creating the boson in the laboratory as an excitation of the all-penetrating Higgs field and studying its properties to understand the origin of this field and its connection to the rest of our Universe. It is especially important because the SM appears to be incomplete, as it cannot describe several puzzles in our Universe, such as the mysteries of dark matter and energy, the apparent lack of antimatter compared to matter, and the hierarchy of masses of all fundamental particles, including the mass of the Higgs boson itself. This may imply that the SM is only an effective field theory (EFT), an approximation of a more general theory. We cannot resolve the full theory yet because this would require us to reach a much smaller linear scale or much larger energy scale in our measurements, not reachable directly at the LHC. However, very important effects, such as hints of new physics, could be observed in the properties of the Higgs boson and how it interacts with the rest of the matter and energy in the Universe within the SM EFT. The Higgs field may become the connection of our known world to the dark, or unknown, world.

Therefore, optimal studies of the Higgs boson's interactions with the known matter and force particles are essential to reveal possible new phenomena. This thesis by Dr. Roskes develops fundamental ideas and advanced techniques for these sophisticated studies. The unique feature of this work is a simultaneous measurement of the Higgs boson's associated production (its "context") and its decay (its "end," in the words of the thesis title) while allowing for multiple parameters sensitive to new phenomena. This includes computer simulation with Monte Carlo techniques of the complicated structure of the Higgs boson interactions (JHU generator); the matrix-element calculation of per-event likelihoods (MELA package) for optimal observables; and advanced fitting methods with hundreds of intricate components that cover all possible parameters and quantum mechanical interference. Optimization of the CMS detector's silicon-based tracking system, essential for these measurements, is also described. The development of the track-based alignment algorithm of the silicon tracking detector (HipPy) and the methods and tools for its validation are discussed in detail. The thesis culminates with a presentation of both phenomenological development and experimental application of the most advanced analysis of LHC data to date in the multiparameter approach to Higgs physics in its single golden four-lepton decay channel. The couplings of the Higgs boson to both vector bosons and fermions are analyzed under the framework of the SM EFT and constraints on quantum properties of the Higgs boson are set. The presented results represent the original work of Dr. Roskes, performed in collaboration with CMS and theory colleagues.

Baltimore, MD, USA Andrei V. Gritsan

Acknowledgements

There are so many people who contributed to the work described in this thesis.

To my advisor, Andrei Gritsan: I cannot thank you enough for everything you have done over the last 8 years, starting when I was a student in your Special Relativity and Waves class and continuing to the present. From the beginning, you have pushed me to 125% of my potential. When the results came out, you always supported me, whether the final result was 80%, 100%, or 110%, and even if the result looked different or took longer than expected. Nothing here would have happened without your vision and your constant support.

Yanyan Gao and Nhan Tran: although I did not get to work with you personally, my work here is a direct extension of the developments you pioneered during your time at JHU and could not have happened without that baseline.

Andrew Whitbeck: thank you for everything you did for me when you were a senior grad student and I was just starting out in research. I particularly appreciated your patience with me and your willingness to always help me learn and answer my questions, even when you had more important things to focus on (like, you know, discovering the Higgs boson).

Sara Bolognesi, Ian Anderson, Chris Martin, Candice You: thank you for working with me during the beginning of my time at JHU. As I was starting research, I learned a lot from each of you.

Meng Xiao: it was a privilege to work with you over the last 6 years. I have learned so much from you and especially from your ability to produce complicated results in a short time while always staying organized and positive.

Ulascan Sarica: thank you for working with me on so many projects over the last 6 years. We spent so much time working together, and we ended up producing amazing results.

Savvas Kyriacou: I enjoyed getting to know you over the last few months and wish you success in continuing these projects.

Markus Schulze: your effort is really the driving force behind JHUGEN and MELA. None of these projects could have happened without your continuous support over the years.

Yaofu Zhou: thank you for your contributions to JHUGEN and MELA, particularly in VH, and for your help with theory questions.

Maggie Eminizer: thanks for always being there to bounce ideas back and forth and give me a boost when things got frustrating. It was great to have someone to talk to who understood the general principles and the CMS software with all its ups and downs, but who also was not working on exactly the same thing and could listen with a fresh ear.

Barry Blumenfeld, Morris Swartz, Petar Maksimovic, and past and present HEP students at JHU: thank you for forming a wonderful environment where we could all collaborate on many different but interrelated aspects of CMS research.

Jered McInerney, Wenzer Qin, Kyle Sullivan, Jason Fan, Victor de Havenon, and Jared Feingold: I really enjoyed working with you as you started research. Wishing you all the best success in the future.

CJLST (CMS Joint 4Lepton Study Team) friends: it was a pleasure working together in such a strong and collaborative group. Thank you for contributing all the pieces needed for $\mathrm{H} \rightarrow \mathrm{VV} \rightarrow 4\ell$ reconstruction. I cannot imagine what would have happened if we would each had to put all the pieces together for each analysis individually.

Tracker alignment friends: I really enjoyed working with you over the last 8 years on CMS. Alignment was my first introduction to research, and I will always value that experience. Special thanks to Patrick Connor, with whom I had the honor to serve as convener.

To everyone else in the CMS collaboration, the LHC team, and the physics department at JHU: your background support, 99% of which I do not even know about, makes all of this possible, and I greatly appreciate everything you have done.

Aleksey Kondratyev, Tianran Chen, and Amitabh Basu: thank you for providing crucial input in preparing templates for the multiparameter analysis. This was the final result I wanted to contribute for this thesis, and it could not have come together without your help.

Andrei Gritsan (again), Morris Swartz (again), Alex Szalay, Amitabh Basu (again), and Tyrel McQueen: thank you for reviewing my thesis and presentation. I was glad to have the opportunity to share the results with you.

And, finally, to my family, Mommy, Daddy, Nomi, and Lani: every single day I spent in grad school was made possible by your constant love and unconditional support. At the best times, you were always there for me and proud of me. At the difficult times, you always knew exactly what I should do next. And at all times, you did everything you could to give me the opportunity to focus on my work without distractions. Wherever I go in life, it will be because I have such a wonderful, loving, supportive family behind me.

As I make the formatting modifications to my thesis in preparation for the publication by Springer, I want to acknowledge one other person, who entered my life just a week before my defense. Miriam, you have made my life absolutely wonderful in the months since then and supported me throughout the process of finishing up my analysis and editing my thesis. Cannot wait to be right there with you through grad school, to your defense, and beyond. I love you.

תושלב"ע

Rabbi Yishmael says: The Torah can be derived through thirteen principles…

12. A matter can be learned from its context, and a matter can be learned from its end…

—*Introduction to Sifra [1], third-century legal commentary on Leviticus*

In enumerating the principles listed here, use this rule: any two principles that follow the same logic are counted together as one principle.
—*Commentary of Ra'avad (1125–1198) to this passage [2]*

1. "Sifra," in *Sifra devei Rav*, A. Shoshana, Ed., vol. 1, Jerusalem: Machon Ofek, 1991, pp. 1–14, ISBN: 9781881255093.
2. A. ben David, "Commentary on Torat Kohanim," in *Sifra devei Rav*, A. Shoshana, Ed., vol. 1, Jerusalem: Machon Ofek, 1991, pp. 15–50, ISBN: 9781881255093. xii

Parts of This Thesis Have Been Published in the Following Journal Articles

1. W. Adam *et al.*, "Alignment of the CMS Silicon Strip Tracker during standalone Commissioning," *JINST*, vol. 4, T07001, 2009. https://doi.org/10.1088/1748-0221/4/07/T07001. arXiv:0904.1220 [physics.ins-det].
2. S. Chatrchyan *et al.*, "Alignment of the CMS Silicon Tracker during Commissioning with Cosmic Rays," *JINST*, vol. 5, T03009, 2010. https://doi.org/10.1088/1748-0221/5/03/T03009. arXiv:0910.2505 [physics.ins-det].
3. The LHC Higgs Cross Section Working Group, "Handbook of LHC Higgs cross sections: 3. Higgs properties," 2013. https://doi.org/10.5170/CERN-2013-004. arXiv: 1307.1347.
4. M. Teklishyn, "Measurement of the η_c (1S) production cross-section via the decay η_c to proton-antiproton final state," PhD thesis, Universite Paris Sud - Paris XI, Sep. 2014.
5. V. Khachatryan *et al.*, "Constraints on the spin-parity and anomalous HVV couplings of the Higgs boson in proton collisions at 7 and 8 TeV," *Phys. Rev. D*, vol. 92, p. 012 004, 2015. https://doi.org/10.1103/PhysRevD.92.012004. arXiv: 1411.3441 [hep-ex].
6. J. Ellis, M. K. Gaillard, and D. V. Nanopoulos, "A Historical Profile of the Higgs Boson," L. Maiani and L. Rolandi, Eds., pp. 255–274, 2016. https://doi.org/10.1142/9789814733519_0014. arXiv:1504.07217 [hep-ph].
7. D. de Florian, C. Grojean, *et al.*, "Handbook of LHC Higgs cross sections: 4. deciphering the nature of the Higgs sector," 2016. https://doi.org/10.23731/CYRM-2017-002. arXiv:1610.07922.
8. A. M. Sirunyan *et al.*, "Measurements of properties of the Higgs boson decaying into the four-lepton final state in pp collisions at $\sqrt{s} = 13$ TeV," *JHEP*, vol. 11, p. 047, 2017. https://doi.org/10.1007/JHEP11(2017)047. arXiv:1706.09936 [hep-ex].
9. "Measurements of properties of the Higgs boson in the four-lepton final state at $\sqrt{s} = 13$ TeV," CERN, Geneva, Tech. Rep. CMS-PAS-HIG-18-001, 2018. [Online]. Available: https://cds.cern.ch/record/2621419.

10. A. M. Sirunyan *et al.*, "Search for a new scalar resonance decaying to a pair of Z bosons in proton-proton collisions at $\sqrt{s} = 13$ TeV," *JHEP*, vol. 06, p. 127, 2018, [Erratum: JHEP 03, 128 (2019)]. https://doi.org/10.1007/JHEP06(2018)127. arXiv:1804.01939 [hep-ex].
11. CMS Collaboration, "Measurements of properties of the Higgs boson in the four-lepton final state in proton-proton collisions at $\sqrt{s} = 13$ TeV," CERN, Geneva, Tech. Rep. CMS-PAS-HIG-19-001, 2019. [Online]. Available: https://cds.cern.ch/record/2668684.
12. A. M. Sirunyan *et al.*, "Constraints on anomalous *HVV* couplings from the production of Higgs bosons decaying to τ lepton pairs," *Phys. Rev. D*, vol. 100, no. 11, p. 112 002, 2019. https://doi.org/10.1103/PhysRevD.100.112002. arXiv: 1903.06973 [hep-ex].
13. A. M. Sirunyan *et al.*, "Measurements of the Higgs boson width and anomalous *HVV* couplings from on-shell and off-shell production in the four-lepton final state," *Phys. Rev. D*, vol. 99, no. 11, p. 112 003, 2019. https://doi.org/10.1103/PhysRevD.99.112003. arXiv:1901.00174 [hep-ex].
14. A. V. Gritsan, J. Roskes, U. Sarica, M. Schulze, M. Xiao, and Y. Zhou, "New features in the JHU generator framework: constraining Higgs boson properties from on-shell and off-shell production," *Phys. Rev. D*, vol. 102, no. 5, p. 056022, 2020. https://doi.org/10.1103/PhysRevD.102.056022.arXiv:2002.09888 [hep-ph].
15. CMS collaboration, Constraints on anomalous Higgs boson couplings to vector bosons and fermions in production and decay $H \rightarrow 4l$channel. Report No.: CMS-PAS-HIG-19-009, to be submitted to *Phys. Rev. D*. http://cds.cern.ch/record/2725543.

Contents

Chapter 1
Introduction

This thesis will describe several analyses measuring the properties of the Higgs boson. As this chapter will describe, the Higgs boson was predicted to exist in 1967 as part of a theoretical explanation of the weak nuclear force. The theory that is now known as the Standard Model of particle physics, or the "SM," was built over the next 20 years, and it describes almost all known particles and interactions. Although the SM is not a *complete* description of the universe—its most obvious omission is gravity—almost all measurements of processes the SM covers agree with the theoretical predictions to incredible precision, with some especially precise measurements reaching 9 digits of agreement.

For a long time, the Higgs boson was the only elementary particle that was predicted by the SM but not observed. After almost 50 years of searching by building experiments designed to probe higher and higher energies, it was finally discovered by the CMS and ATLAS collaborations in 2012. As the most recently discovered fundamental particle, the Higgs boson is a natural target for higher-precision experimental tests of the SM's predictions. If deviations from the SM are detected, they would be a hint to further beyond SM, or "BSM," physics, such as new particles or interactions, and possibly open doors to understanding other mysteries of physics, such as the identity of dark matter.

1.1 Elementary Particles

Philosophical arguments for the idea that matter can only be divided up to a certain fundamental limit, known as an atom, go back to the ancient Greeks. Ultimately, the word "atom" ended up with a slightly different connotation than the Greek philosophers intended. The objects now known as atoms, although they form a useful basic unit in describing how matter interacts under many normal circumstances, are made of protons, neutrons, and electrons. The electron, isolated

J. Roskes, *A Boson Learned from its Context, and a Boson Learned from its End*, Springer Theses, https://doi.org/10.1007/978-3-030-58011-7_1

by J.J. Thomson in 1897, was the first particle discovered that, as far as we know, is actually an elementary particle. A few years later, in the results that are considered to be the birth of quantum mechanics, Planck [1] and Einstein [2] showed that light is quantized in packets of energy, which came to be known as photons.

With these two discoveries, we have examples of each of the two types of elementary particles: fermions, which have half-integer spin, obey the Pauli exclusion principle, and therefore make up matter; and bosons, which have integer spin and transmit forces between fermions.

Further developments in particle physics over the next few decades came through experiments probing atomic structure. Rutherford proposed in 1911, based on the famous gold foil experiments, that an atom's positive charge is concentrated in the central nucleus [3], and his subsequent experiments revealed that the nucleus contains protons [4]. The existence of neutrons was suspected for years and finally confirmed experimentally in 1932 [5]. Around the same time, to reconcile the energy spectrum of β-decay with conservation of energy, Pauli proposed the existence of another very light, neutral particle, known as the neutrino [6].

To explain all of these observations, two new interactions were required. The weak force makes β decay possible by allowing neutrons to change into protons. However, calculations showed that this force is too weak to explain how the nucleus, with many positively-charged protons in close proximity, remains bound. Yukawa [7] proposed an additional force, the strong nuclear force, mediated by a particle with a mass between the masses of the proton and the electron. The first particle discovered that fit this criterion was the muon, which surprised everybody because it has nothing to do with the strong force and is actually a lepton—except for its mass, it behaves exactly like the electron. The pion, which really is the particle predicted by Yukawa, was discovered in 1947 [8, 9]. Over the next decades, a large number of other strongly-interacting bosons, known as mesons, and fermions, known as baryons, were discovered. Collectively, these strongly interacting particles are known as hadrons. (Yukawa's description of the strong nuclear force should not be confused with the later development of the quark model and quantum chromodynamics, which will be described in Sect. 1.5.)

1.2 Gauge Symmetry

In the SM's current form, its forces are all derived from gauge symmetries of the interacting particles. The electromagnetic force, mediated by the photon A_μ, is generated by a $U(1)$ gauge symmetry α of the charged fermion fields ψ: its Lagrangian

$$\mathcal{L} = \bar{\psi}(i\gamma^\mu D_\mu - m)\psi - \frac{1}{4}F_{\mu\nu}^2, \tag{1.1}$$

where $F_{\mu\nu} = \partial_\mu A_\nu - \partial_\nu A_\mu$, $D_\mu = \partial_\mu - iqA_\mu$, q and m are the fermion's charge and mass, and γ^μ are the Dirac matrices, is invariant under the transformation

$$\begin{aligned} \psi &\rightarrow e^{-i\alpha(x)}\psi \\ A_\mu &\rightarrow A_\mu + \frac{i}{q}\partial_\mu\alpha(x) \end{aligned} \tag{1.2}$$

Since the photon is massless, everything works well.

However, trying to apply the same procedure to a short-range force leads to trouble. If a force has short range, this implies that the particle that transmits has mass. A mass term in the Lagrangian, proportional to $A_\mu A^\mu$, would violate this gauge symmetry.

Explanations for the weak force, which is a short-range force, face this difficulty. On the other hand, because the weak force involves charged currents, which turn electrons into neutrinos or protons into neutrons, the particles that transmit it, now known as W bosons, must be charged, meaning that its theory has to be combined with the electromagnetic force in some way. In 1959, Glashow, Salam, and Ward developed a theory of the electroweak interaction [10, 11], but the difficulty of assigning mass to the W and Z bosons, and hence limiting the range of the force, remained.

1.3 Symmetry Breaking

In the meantime, an early attempt to understand the nature of the strong force was undertaken by Yochiro Nambu in 1960 [12, 13] by analogy with a similar feature in superconductivity. In the ground state of a superconductor, all of the electrons in the material form correlated Cooper pairs. The underlying theory describing how the electrons interact with each other and with the nuclei is the electromagnetic force, which respects gauge symmetry. However, the ground state of the system involves correlated electrons, which requires a fixed relative phase. Although there is no preference for any particular phase, once a phase is chosen at the time the material becomes a superconductor, it is fixed. As will be described later in the context of the Higgs potential, this "spontaneous symmetry breaking" implies the existence of massless bosons that transmit interactions between excitations from the ground state. In a superconductor, excitations take the form of linear combinations between electrons and holes.

Similarly to Cooper pairs of electrons in a superconductor, Nambu suggested that protons and neutrons are mixture states between more fundamental particles, one left-handed and one right-handed. Nucleon interactions are invariant under rotations between left- and right-handed states; however, the nucleon's mass term breaks this symmetry. Therefore, as in the superconductor, there should exist a massless boson that transmits a force between nucleons. Nambu identified this boson with the pion, and suggested that the underlying theory of nucleons, unknown at the time, does not exactly respect chiral symmetry, making the pion mass small but nonzero. Using this model, Nambu was able to explain several features of pions and heavier mesons.

1.4 The Higgs Mechanism

Nambu's result was generalized over the next few years by Goldstone, Weinberg, and Salam [14, 15]. The result is that any spontaneously broken symmetry generates a massless particle, known as a Goldstone boson.

In simultaneous papers by Brout and Englert [17], Higgs [18, 19], and Guralnik et al. [20] in 1964, it was shown that spontaneous symmetry breaking can result in massive gauge bosons, which would be necessary to describe the short-range weak force as a gauge theory. They considered a complex scalar field ϕ with interactions following the Mexican hat potential parameterized by constants λ and μ,

$$V(\phi) = \frac{1}{4}\lambda^2|\phi|^4 - \frac{1}{2}\mu^2|\phi|^2, \tag{1.3}$$

illustrated in Fig. 1.1. Before any interactions with other fields are included, the corresponding Lagrangian is

$$\mathcal{L} = \frac{1}{2}|d_\mu\phi|^2 - \frac{1}{4}\lambda^2|\phi|^4 + \frac{1}{2}\mu^2|\phi|^2 \tag{1.4}$$

The vacuum, or ground state, is where the energy is minimized: a circle at $\phi^\dagger\phi = \left(\frac{\mu}{\lambda}\right)^2$. Again, the symmetry of the Lagrangian, which was invariant under $\phi \to e^{i\alpha}\phi$, is broken by choosing one of these ground states.

Rewriting the Lagrangian by starting from one of these ground states, we can set

$$\phi = v + h + i\xi,$$

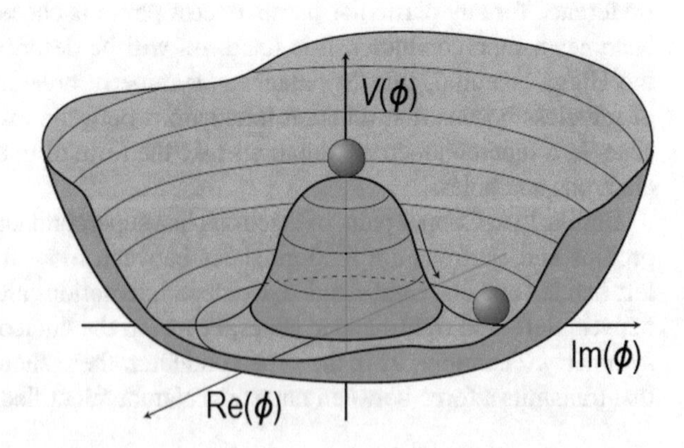

Fig. 1.1 Plot of the Mexican hat potential, Eq. (1.3). This plot is from [16]

where

$$v = \frac{\mu}{\lambda}$$

and h and ξ are real scalar fields. Substituting this into Eq. (1.4), we find, among other terms,

$$\frac{1}{2}(d_\mu h)^2 - \mu^2 h^2 + \frac{1}{2}(d_\mu \xi)^2. \tag{1.5}$$

Of two particles in this parameterization, one, ξ, is massless as expected for a Goldstone boson, and the other, h, has a mass of $\sqrt{2}\mu$. The other terms are cubic or quartic in the fields and give various interactions of the form hhh, $h\xi\xi$, $hhhh$, $\xi\xi\xi\xi$, and $hh\xi\xi$, with coupling strengths that are functions of μ and λ.

If the original field Φ couples to a gauge field A_μ, then the rewritten Lagrangian, derived from Eq. (1.4) with $d_\mu \rightarrow D_\mu$, will also contain terms that look like $A_\mu d^\mu \xi$. This term means that A_μ can turn into ξ and vice versa. By further reparameterization by choosing a particular gauge, ξ can be removed from the equation entirely. The $A_\mu d^\mu \xi$ terms become a mass term for A_μ. We now have a way to introduce massive gauge bosons.

In Peter Higgs' paper, he pointed out an important consequence of this model: the scalar field h should be observable as a particle. Although he personally downplays his role and prefers to share credit with the others who independently discovered this possibility, this type of particle is generally known as a Higgs boson.

Although fermions are allowed to have a mass term without affecting gauge invariance, a coupling between fermions and Φ would also naturally introduce a mass term proportional to the coupling strength as well as couplings to h and A_μ. From experiments so far, it seems to be the case that the Higgs mechanism also generates fermion masses.

Using the Higgs mechanism and building on Glashow and Salam's earlier work, the full theory of the weak force and its combination with electromagnetism came through Weinberg and Salam in 1967 [21, 22]. Because there needed to be three massive vector bosons, the full electroweak theory involves a complex scalar doublet instead of a single complex scalar as described here. Three of the four degrees of freedom get "eaten" to form the masses of the Z and W bosons, and the fourth is left as a Higgs boson.

Electroweak theory, combined with earlier experiments revealing how the weak force affects leptons and baryons, predicted properties of the Z and W bosons. The first interaction mediated by the Z boson, neutrino scattering, was observed in 1973, and the Z and W bosons themselves were observed in 1983.

1.5 The Strong Force and QCD

Around the same time as the weak force, the strong force also came to be understood better. Gell-Mann [23] and Zweig [24, 25] predicted in 1964 that baryons and mesons are made of one of three types of more fundamental particles, known as the up, down, and strange quarks. This theory explained the structure and interactions of the many, many types of baryons and mesons in existence. Each quark carries a "color charge," and the differently colored quarks can be rotated into each other by an $SU(3)$ symmetry, similar to the way the simpler electric charge and electromagnetic force are invariant under a $U(1)$ symmetry.

The fundamental strong force, known as quantum chromodynamics or QCD, is transmitted between these quarks by eight gluons. The strong nuclear force, which affects baryons and mesons and is responsible for holding nuclei together, comes from interactions between the quarks and gluons that make up the nucleons involved. Because gluons are massless, there is no need for a Higgs mechanism here.

The strong force is observed to only act at short distances for a different reason. Unlike the electromagnetic force, the mathematical description of QCD behaves well at high energy and badly at low energy. At the shortest distances, quarks do not interact, and this is known as "asymptotic freedom". However, when quarks move too far away from each other, the interaction energy increases significantly. The result is "color confinement": free particles with color charge do not exist. Instead, when a process with high enough energy splits a baryon or meson, more quark-antiquark pairs are produced from the vacuum to create more baryons and mesons, all of which are colorless and do not directly interact via the strong force.

This presents experimental advantages and disadvantages. In a high energy process, QCD effects are small; they can be expanded around the lowest order contribution and in many cases can be neglected entirely. On the other hand, when a process produces quarks or gluons, they split into many more quarks and gluons. Many more quark-antiquark pairs are produced to neutralize the remaining color charge. The result is a jet containing baryons and mesons, which is more complicated to detect and measure than electrons or muons.

The existence of a fourth quark, known as charm, was proposed shortly after the quark model itself, and in 1970 Glashow, Iliopoulos, and Maiani used the possibility to explain why flavor changing neutral currents, which would involve strange quarks decaying to down quarks, are suppressed [26]. Once it was discovered in the form of the J/ψ meson in 1974, there were four known quarks, up, down, charm, and strange, corresponding to four leptons, the electron and its neutrino and the muon and its neutrino.

1.6 The Standard Model

The combined description of the electromagnetic and strong forces was developed in the 1970s and became known as the Standard Model. Since its initial inception, an additional generation of quarks and leptons was proposed and discovered, consisting of the tau lepton and an associated neutrino and the top and bottom quarks.

The resulting theory involves 61 fundamental particles. There are six flavors of quarks, each of which comes in three colors, and six leptons. Each of these particles also has a corresponding antiparticle, so the total comes to 48 fermions. There are 12 vector bosons, which transmit forces: 8 gluons for the strong force, the $W^{\pm}$ and Z bosons for the weak force, and the photon for the electromagnetic force. Finally, the scalar Higgs boson brings the total to 61.

The quarks, which have color charge, interact through the strong force, mediated by the octet of gluons, which also interact among themselves. The quarks and charged leptons, which have electric charge, interact through the electromagnetic force, mediated by the photon. Finally, all types of fermions interact through the weak force, mediated by the Z and W bosons. The Z and W bosons also interact with each other, and the W boson, which has electric charge, interacts with the photon. The relationships and interactions between the SM particles are summarized in Fig. 1.2.

1.7 Limitations of the Standard Model

As mentioned in the introduction to this chapter, almost all experiments probing the SM find agreement with the theoretical predictions, although there are a few outliers

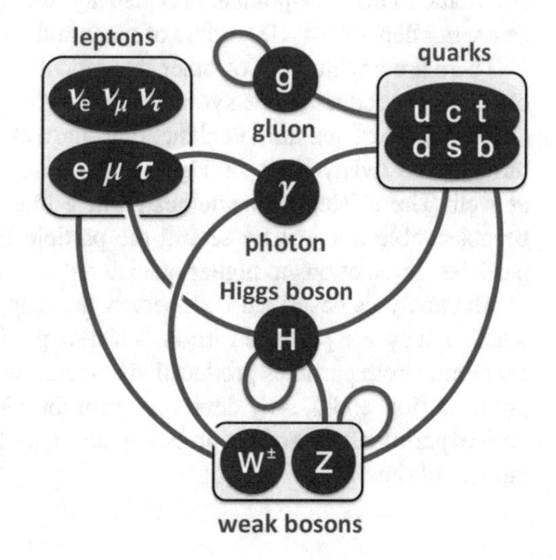

Fig. 1.2 Particles of the Standard Model, shown in the black circles and ovals, and their interactions with each other, shown as blue lines connecting them. Lines looping from a particle back to itself indicate self-interactions. This clever illustration is from [27]

under further investigation, such as the muon's anomalous magnetic moment. However, many phenomena are left unexplained by the SM. This is a list of a few important holes that may be better understood by examining the Higgs boson, which can reveal properties of the Higgs field.

From astrophysics, we know that dark matter exists, and observations place limits on how strongly or weakly it can interact with regular matter. No experiment has managed to detect dark matter particles on Earth other than through their gravitational influence on astrophysics and cosmology. If dark matter's mass is generated through interactions with the Higgs field, then it should also interact with the Higgs boson.

The universe contains much more matter than antimatter, which can again be determined from astrophysical observations. Electromagnetism, the strong force, and gravity treat matter and antimatter identically. The weak force does interact differently with matter than with antimatter, violating CP symmetry. However, calculations show that the small magnitude of CP violation in the weak force is insufficient to produce the huge ratio of matter to antimatter that we observe. The SM predicts that the Higgs boson is CP even. If this predication fails and a CP odd component is present as well, the Higgs boson's interactions would be an additional source of CP violation.

Another mystery of the Standard Model is the hierarchy problem. Quantum corrections to the Higgs boson's mass, generated by loop interactions with other particles, are of order 10^{19} GeV. This means that to produce the value that we observe, the "bare" mass, before those corrections, must be exactly $(x \times 10^{19} + 125)$ GeV. There is no known mechanism for this fine-tuning. Several theories exist that would explain it, and predict new particles and/or interactions as part of that explanation.

A similar puzzle relates to dark energy, which accelerates the expansion of the universe. All of the fields of the Standard Model, including the Higgs field, should contribute to the zero-point energy density of empty space, but calculations result in an energy density 50–100 orders of magnitude larger than what is observed.

There are any number of other theoretical reasons to suggest that new particles, breaking or producing one symmetry or another, might exist.

The nature of quantum field theory is that a change in the behavior of one particle, through the twisty lines in Fig. 1.2, has effects on all other particles' behavior as well. The existence of another particle that interacts with SM particles would be observable not just by seeing the particle itself, but also by measuring other particles' interactions to higher precision.

The analyses described here search for changes to the Higgs boson's behavior, whether they are produced through a new particle or some other mechanism, by examining both particles produced in association with it (its "context") and its decay products (its "end"). Any deviation from the SM predictions would be interesting both experimentally and theoretically and may help to explain one or more of the puzzles of the SM.

References

1. M. Planck, Ueber das gesetz der energieverteilung im normalspectrum. Ann. Phys. **309**(3), 553–563 (1901). https://doi.org/10.1002/andp.19013090310
2. A. Einstein, Concerning an heuristic point of view toward the emission and transformation of light. Ann. Phys. **17**, 132–148 (1905)
3. E. Rutherford, The scattering of alpha and beta particles by matter and the structure of the atom. Philos. Mag. Ser. 6 **21**, 669–688 (1911). https://doi.org/10.1080/14786440508637080
4. E. Rutherford, Collision of α particles with light atoms. IV. An anomalous effect in nitrogen. Philos. Mag. Ser. 6 **37**, 581–587 (1919). https://doi.org/10.1080/14786431003659230
5. J. Chadwick, Possible existence of a neutron. Nature **129**(3252), 312 (1932). https://doi.org/10.1038/129312a0
6. W. Pauli, Rapports du septieme conseil de physique solvay, Brussels (Gauthier Villars, Paris, 1934) (1933)
7. H. Yukawa, On the interaction of elementary particles I. Proc. Phys. Math. Soc. Jpn. **17**, 48–57 (1935) [Prog. Theor. Phys. Suppl. 1, 1 (1935). https://doi.org/10.1143/PTPS.1.1]
8. G.P.S. Occhialini, C.F. Powell, Nuclear disintegration produced by slow charged particles of small mass, in *Bristol 1987, Proceedings, 40 Years of Particle Physics* [Nature **159**, 186 (1947)]. https://doi.org/10.1038/159186a0
9. D.H. Perkins, Nuclear disintegration by meson capture. Nature **159**, 126–127 (1947). https://doi.org/10.1038/159126a0
10. S.L. Glashow, The renormalizability of vector meson interactions. Nucl. Phys. **10**, 107–117 (1959). https://doi.org/10.1016/0029-5582(59)90196-8
11. A. Salam, J.C. Ward, Weak and electromagnetic interactions. Nuovo Cimento **11**, 568–577 (1959). https://doi.org/10.1007/BF02726525
12. Y. Nambu, G. Jona-Lasinio, Dynamical model of elementary particles based on an analogy with superconductivity. I. Phys. Rev. **122**, 345–358 (1961). https://doi.org/10.1103/PhysRev.122.345 [127 (1961)]
13. Y. Nambu, G. Jona-Lasinio, Dynamical model of elementary particles based on an analogy with superconductivity. II. Phys. Rev. **124**, 246–254 (1961). https://doi.org/10.1103/PhysRev.124.246 [141 (1961)]
14. J. Goldstone, Field theories with ≪ superconductor ≫ solutions. Il Nuovo Cimento **19**(1), 154–164 (1961). ISSN:1827-6121. https://doi.org/10.1007/BF02812722
15. J. Goldstone, A. Salam, S. Weinberg, Broken symmetries. Phys. Rev. **127**, 965–970 (1962). https://doi.org/10.1103/PhysRev.127.965
16. J. Ellis, M.K. Gaillard, D.V. Nanopoulos, A historical profile of the Higgs boson, in *The Standard Theory of Particle Physics*, ed. by L. Maiani, L. Rolandi (2016), pp. 255–274. https://doi.org/10.1142/9789814733519_0014, arXiv:1504.07217 [hep-ph]
17. F. Englert, R. Brout, Broken symmetry and the mass of gauge vector mesons. Phys. Rev. Lett. **13**, 321 (1964). https://doi.org/10.1103/PhysRevLett.13.321
18. P.W. Higgs, Broken symmetries, massless particles and gauge fields. Phys. Lett. **12**, 132 (1964). https://doi.org/10.1016/0031-9163(64)91136-9
19. P.W. Higgs, Broken symmetry and the mass of gauge vector mesons. Phys. Rev. Lett. **13**, 508 (1964). https://doi.org/10.1103/PhysRevLett.13.508
20. G. Guralnik, C. Hagen, T. Kibble, Global conservation laws and massless particles. Phys. Rev. Lett. **13**, 585 (1964). https://doi.org/10.1103/PhysRevLett.13.585
21. S. Weinberg, A model of leptons. Phys. Rev. Lett. **19**, 1264–1266 (1967). https://doi.org/10.1103/PhysRevLett.19.1264
22. A. Salam, Weak and electromagnetic interactions, in *Conference Proceedings*, vol. C 680519 (1968), pp. 367–377
23. M. Gell-Mann, A schematic model of baryons and mesons. Phys. Lett. **8**(3), 214–215 (1964). ISSN:0031-9163. https://doi.org/10.1016/S0031-9163(64)92001-3
24. G. Zweig, An SU(3) model for strong interaction symmetry and its breaking. Version 1 (1964)

25. G. Zweig, An SU(3) model for strong interaction symmetry and its breaking. Version 2, in *Developments in the Quark Theory of Hadrons: Vol. 1, 1964–1978*, ed. by D. Lichtenberg, S.P. Rosen (1964), pp. 22–101
26. S.L. Glashow, J. Iliopoulos, L. Maiani, Weak interactions with lepton-hadron symmetry. Phys. Rev. D **2**, 1285–1292 (1970). https://doi.org/10.1103/PhysRevD.2.1285
27. M. Teklishyn, Measurement of the $\eta_c(1S)$ production cross-section via the decay η_c to proton-antiproton final state. PhD thesis, Universite Paris Sud - Paris XI, Sept 2014

Chapter 2
The Experiment

The heaviest particle of the Standard Model is the top quark at 173 GeV, followed by the H (125 GeV), Z (91.2 GeV), and W (80.4 GeV) bosons. To produce these particles, you need to give the system at least enough energy to cover the mass. In addition, for a given targeted process, only a small fraction of collisions will actually produce that process, so a large number of collisions, high *luminosity*, is needed to accumulate enough events to draw interesting conclusions. Finally, you need a detector capable of measuring the type, energy, and momentum of the particles produced.

Astrophysical events often involve enormous amounts of energy, and the most energetic particles detected have come from space. However, this kind of data is not useful for directly measuring properties of the Higgs boson, which requires detecting multiple decay products.

The Large Hadron Collider (LHC) is designed to cover the first two requirements, energy and luminosity. The collisions of the LHC are at the highest energy and luminosity of any collider to date. The Compact Muon Solenoid (CMS) experiment is designed to measure particle decays to high precision. Although they were also designed to be capable of many different analyses, one of the primary design goals of the LHC and its detectors was to produce, discover, and characterize the Higgs boson.

2.1 The Large Hadron Collider

The Large Hadron Collider is located 100 m under the French and Swiss countryside near Geneva, shown in Fig. 2.1. It accelerates protons around a 27 km tunnel and collides them together. Several smaller accelerators are used to accelerate the protons to successively higher energies until they are injected from the Super Proton Synchrotron (SPS, shown in blue in the picture) into the LHC at an energy of

J. Roskes, *A Boson Learned from its Context, and a Boson Learned from its End*, Springer Theses, https://doi.org/10.1007/978-3-030-58011-7_2

Fig. 2.1 An aerial picture of the ground above the LHC, with its path shown in a circle and the interaction points labeled with the experiments located there. This picture is from [1]

450 GeV. All around the tunnel, a series of 1232 dipole magnets direct the protons through the curves of the tunnel, and higher order magnets focus the beams. The protons circulate in two beams, traveling in opposite directions. Once they are accelerated to the final energy, currently 6.5 TeV, the beams are brought together to collide at four interaction points, with a center of mass energy of 13 TeV.

Each interaction point houses one of the four large LHC experiments: ATLAS, CMS, LHCb, and ALICE. ATLAS and CMS are both general-purpose detectors, designed to be sensitive to multiple types of possible new physics. LHCb is optimized for measurements in quark flavor physics, particularly for precise measurements of hadrons containing the b quark. ALICE is designed for the periods of time when the LHC collides ion nuclei together and particularly focuses on studying the quark–gluon plasma that can be formed at the extremely high temperatures generated by the collisions.

During a fill, the LHC collides bunches of protons every 25 ns. Approximately 2800 bunches at a time circulate in the tunnel, and each one contains around 10^{11} protons. Each bunch crossing results in an average of 37–38 collisions, giving a total of 1.5 billion collisions per second. Most of those collisions do not produce anything interesting. In order to calculate the expected number of collisions that result in a particular process, we use the process' *cross section*, measured in units

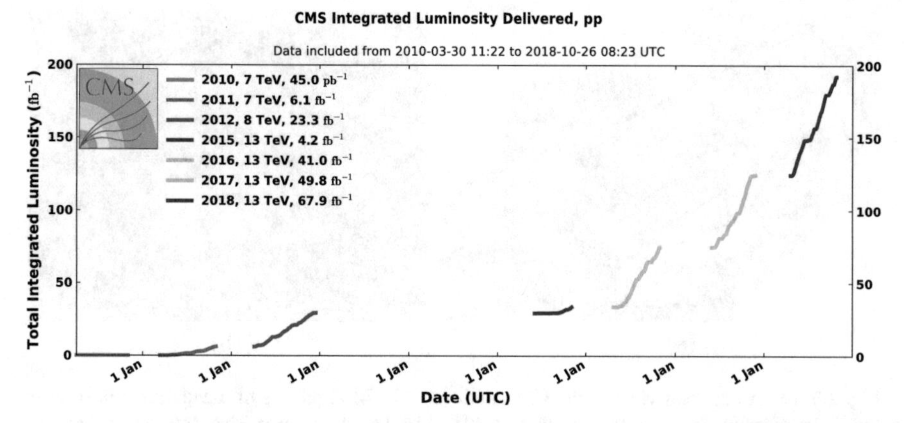

Fig. 2.2 Plots of the integrated luminosity delivered to CMS over the years since the start of LHC operations. This plot is from [2]

of area, and the integrated luminosity, with dimensions of $\frac{1}{\text{area}}$. The cross section is the quantum mechanical equivalent of the cross sectional area of a bucket, and the integrated luminosity is the equivalent of the number of raindrops falling per area. The product of the two gives the number of raindrops that fell into the bucket.

Due to improvements in the LHC over the years, the luminosity delivered has increased, as shown in Fig. 2.2. Additionally, cross sections for processes generally increase at larger energy, giving a further boost to the number of events involving rare processes in the more recent runs of the LHC.

2.2 The CMS Experiment

The Compact Muon Solenoid detector is one of the multi-purpose detectors at the LHC. It is built to cover as much angular space as possible, so that it can detect almost all of the particles produced in collisions. It is named for three defining features of its construction:

- CMS is compact: although it is only a fifth the volume of ATLAS, it weighs almost twice as much.
- CMS is especially optimized for precise measurements of muons, as the next few sections will discuss.
- The 3.8 T solenoid magnet is especially strong in order to increase the precision of momentum measurements of long-lived charged particles.

The detector is arranged in layers, each sensitive to different types of particles. The innermost layer, the tracker, measures the trajectory of all charged particles while having as little effect as possible on their momentum and energy. The electromagnetic calorimeter, or ECAL, measures the energy of electrons and photons,

Fig. 2.3 (**a**) A cut open view of the CMS detector [3]. (**b**) A picture of me standing in front of the CMS detector, taken in the summer of 2013 [4]. The most distinctive parts visible are the interleaved white muon chambers and red return yoke

absorbing them in the process, while having a small effect on hadronic particles and muons. The hadronic calorimeter absorbs hadronic particles and measures *their* energy, while muons are again mostly unaffected. Finally, the muon system provides a second measurement of muons' momentum, and matching to tracks in the tracker identifies those tracks as belonging to muons.

Figure 2.3 shows a cross section of the CMS detector and a picture of me standing in front of it.

2.2.1 Coordinate System

The natural geometry for a collider experiment's detector is cylindrical. There is one preferred axis, the z axis where the protons enter the detector, but the x and y axes are, for the purpose of the collisions, arbitrary. The various parts of the detector are built in two parts: a barrel, to measure particles that travel perpendicular to the beam axis, and an endcap, to measure particles that travel at larger angles. Although the barrel and endcap are of a given subdetector provide the same types of measurements, they are constructed somewhat differently. The endcaps receive a higher flux of particles, which adds two additional constraints to their design: they need to withstand more intense conditions, and they must perform higher granularity measurements in order to distinguish between particles that come at the same time.

The z axis is chosen to be along the beamline. The y axis points up towards the sky, and the x axis is horizontal. Although collisions are azimuthally symmetric, gravity points in the $-y$ direction, which is important when considering the construction of the detector and its movement over time. In addition, cosmic rays, which are used in calibration, typically travel in an almost vertical direction.

We often use the standard azimuthal angle ϕ. While the radial angle θ is used occasionally, particularly when cosmic rays are involved, the more useful quantity for particle collisions is the pseudorapidity:

$$\eta = -\ln \tan \frac{\theta}{2} \tag{2.1}$$

Pseudorapidity transforms more nicely than θ under Lorentz boosts in the z direction, which are used frequently in order to work in the center-of-mass frame of a collision. In addition, the number of particles produced in between η and $\eta + \Delta\eta$ is approximately constant as a function of η.

2.2.2 *Magnet*

The CMS magnet surrounds the inner three parts of the detector. Inside the magnet, it produces an almost uniform magnetic field of 3.8 T, parallel to the beam pipe of the LHC. All particles that have a magnetic moment—electrons, muons, pions, kaons, protons, and neutrons are the ones that live long enough to be appreciably affected—are deflected in the magnetic field, and the curvature of the trajectory is proportional to the particle's momentum. A stronger magnetic field also produces sharper curvature, giving a more precise momentum measurement. As mentioned above, the magnetic field strength was one of the motivations for the design choices in constructing the CMS detector.

The magnetic field is produced by a superconducting solenoid, 12.9 m long and 5.9 m in diameter. It carries a current of 19.5 kA. The iron return yoke of the magnet ensures that the return flux of the magnetic field goes through the muon chambers, so that muon tracks curve and their measurements carry momentum information to complement the information from the tracker.

2.2.3 *Parts of the CMS Detector*

2.2.3.1 Silicon Tracker

The silicon tracker is the innermost part of the detector and provides precise position and momentum measurements for all kinds of charged particles. As a charged particle travels through the tracker volume, it passes through the silicon sensors and excites some of the electrons in the silicon. The detectors collect that charge and provide micron-level resolution of the position of the hit. By matching hits, CMS reconstructs a track and determines the trajectory of the particle.

In addition to a curvature measurement that determines the momentum, the tracker also provides *vertexing* to determine where particles originate. Because each bunch crossing involves an average of around 38 collisions, it is important to determine which particles come from the collision of interest (the primary vertex) and which come from other collisions (pileup). Additionally, certain particles, like b quarks and τ leptons, live long enough to travel a short distance before decaying, producing a secondary vertex. By tracing the tracks back to their origin, we can determine each track's point of origin and associate it to the correct vertex.

The tracker is made of 17,004 silicon modules, rectangles measuring a few centimeters on a side. It is comprised of three several shells, each of which takes the form of a barrel and two endcaps in order to cover a large solid angle and capture as many of the particles produced in collisions as possible. The modules of the innermost shell, the barrel pixel (BPIX) and two forward pixel (FPIX) endcaps, which receive the most hits and require the highest precision, are contain pixels of 100 μm by 150 μm. A voltage is applied across the module, so that when a particle passes through a pixel module and creates electron-hole pairs in the silicon atoms, the charges move to the front of the module and are collected by a readout chip. Although the charge collection is done by individual pixels, the silicon module is made of one connected piece, and the charge deposited near one pixel is collected not only in that pixel but also in its neighbors. The readout chip measures the distribution of charge across pixels, and from this shape CMS can determine the position of the hit to a precision of several microns, more than an order of magnitude smaller than the size of the pixels themselves.

The middle shell is made of the Tracker Inner Barrel (TIB) and Disk (TID), and the outer shell consists of the Tracker Outer Barrel (TOB) and End Cap (TEC). Because these layers receive fewer hits per unit area than the innermost shell, the charge is collected by strips instead of pixels. While the width of each strip is around 100 μm like the pixels, the length is as long as the module, several cm. For this reason, the strip tracker provides only a one-dimensional measurement of the hit position within the module. Some of the strip modules are constructed as stereo modules, with two layers rotated by 100 μrad, or about 6°, with respect to each other. The combination of the two measurements gives some additional information that provides a less precise two-dimensional measurement.

The tracker measures the position of hits within a module, while the quantity of interest is actually the position of the hits in 3D space. While the transformation is easy to make, it introduces an additional possible source of error—our knowledge on the hit's position in space is limited by our knowledge of the module's position in space. The procedure to determine the module positions is discussed in detail in Chap. 3.

The tracker is sensitive to all charged particles, including electrons, muons, and several hadrons. It cannot measure the energy and momentum of neutral particles, including, most notably, photons and neutrinos.

2.2.3.2 Electromagnetic Calorimeter

The electromagnetic calorimeter, or ECAL, measures the energy of electrons and photons. In the case of elections, information from ECAL complements the momentum measurement of the tracker; for photons, ECAL provides the *only* measurement of their energy and momentum. The presence or absence of a track in the tracker also allows CMS to distinguish between electrons and photons.

ECAL's is constructed from 61,200 lead tungstate ($PbWO_4$) crystal scintillators in the barrel region and 7324 more in the endcap. $PbWO_4$ is a dense material with a radiation length of about 8.9 mm, but is also transparent. When an electron or positron passes through these crystals, it recoils off nuclei and produces *bremsstrahlung*, braking radiation. Similarly, in the presence of the heavy nuclei, high energy photons can convert into an electron-positron pair. The resulting cascade is known as an electromagnetic shower.

The low energy photons produced by the shower continue to the end of ECAL, where they are captured and measured. To amplify the signal and obtain a better measurement, photomultipliers in the form of avalanche photodiodes in the barrel region or vacuum phototriodes in the endcap region are used.

An electron or photon's energy decreases, on average, by $1/e$ while traveling a distance in the material equal to its radiation length. Because the width of ECAL is around 25 times the radiation length, electrons and photons deposit virtually all of their energy in ECAL and do not affect the measurements in the rest of the detector. Muons and even the lightest hadrons are significantly heavier than electrons and do not lose a significant amount of energy to bremsstrahlung. They continue to the outer parts of the detector. Neutrinos, as before, are unaffected.

2.2.3.3 Hadronic Calorimeter

The next subdetector, the hadronic calorimeter or HCAL, measures the energy of hadrons. As for electrons, the measurements from charged hadrons can also be matched to tracks in the tracker, while neutral hadrons, such as neutrons, only leave energy deposits in HCAL.

Unlike ECAL, HCAL is designed as a sampling calorimeter: not all of the energy is actually measured. Instead, it is formed from alternating absorber and scintillator layers. The absorber layers are made of brass, with a nuclear interaction length of about 16 cm. On this length scale, hadrons that pass through HCAL interact with the nuclei, lose their energy, and produce more, lower energy hadrons. This phenomenon is similar to the electromagnetic shower in ECAL, but involves interactions between the hadrons and the nuclei and proceeds through the strong force, with pions and other mesons taking the place of photons. The scintillator layers measure the energies of the hadronic particles that pass through them. By measuring this captured energy, HCAL can determine the energy of the original particle.

Like the other parts of the detector, HCAL is divided into barrel and endcap regions. In addition, there is an outer region outside the magnet, placed there to capture any particles that were missed by the barrel, and a forward region, designed to capture radiation that travels almost parallel to the beamline. The width of HCAL in the barrel region is around 10 times the nuclear interaction length, so it captures almost all of the energy of hadronic particles. The only particles left are muons and neutrinos.

2.2.3.4 Muon System

The muon system is designed to give precise measurements of muons. As the only charged particles not absorbed by ECAL and HCAL, muons are typically the only particles that leave a track in the muon system. In the barrel region, drift tube chambers (DTs) are used to detect muons, while in the endcap, cathode strip chambers (CSCs) are used instead. In both the barrel and endcap regions, there are also resistive plate chambers (RPCs). All three types of chambers are filled with gas, and the gas atoms are ionized by muons that pass through. In the DTs, the electrons drift to wires. In the CSCs, both the electrons and ions are collected by arrays of cathode and anode wires. In the RPCs, the electrons are collected by strips. The RPCs are especially precise at determining the time when muons arrive, which is necessary to associate the muon with a particular event, while the DTs and CSCs have better spatial resolution.

2.2.3.5 Neutrinos

Neutrinos are not detected by any part of the CMS detector and escape into space. To measure neutrinos or hypothetical BSM particles that barely interact with normal matter, we use conservation of momentum. Each collision involves protons moving in the z direction, with no momentum in the x or y directions. Therefore, the sum of the products' transverse momentum $\vec{p}_T$ must be zero as well. Any deviation is known as missing transverse energy, or MET, and is a sign of neutrinos or other undetected particles. This is one reason why capturing all of the other particles is so important: any missed particle results in an incorrect MET measurement.

2.2.4 *Trigger*

The LHC provides over a billion collisions per second, and most of those collisions produce nothing interesting. It is impossible to store the data from all of these collisions. Therefore, a solution must be found to distinguish between interesting and useless events, and this must happen fast enough so that the measurements made by the detector can be saved or dropped.

To accomplish this, a two-level triggering system is used. First, the fast level 1 (L1) trigger, using imprecise, raw measurements, determines whether to keep or drop the event. This is all done using custom hardware, and reduces the rate to around 100,000 events per second. Second, events that pass the L1 trigger are given to the high level trigger (HLT), which runs on software and has access to more information about the event. Around 100 events per second pass the second trigger. Events passing both triggers are saved and can be used in analyses.

2.2.5 *Particle Identification and Reconstruction*

To assemble all of the information from the subdetectors into a full picture of a collision event, the particle flow algorithm [6] is used. As mentioned above, each event involves an average of 37–38 collisions, which can all produce particles in overlapping regions of the detector. A method is needed to sort the particles and match information between different parts of the detector. Figure 2.4 summarizes how each type of particle interacts with the various parts of the CMS detector.

The particle flow algorithm first looks for muons, which are the cleanest signal, and matches tracks from the tracker to those from the muon chambers. Those tracks are then removed from consideration, and then the algorithm looks at electrons, matching energy deposits in ECAL with tracks in the tracker. The remaining energy

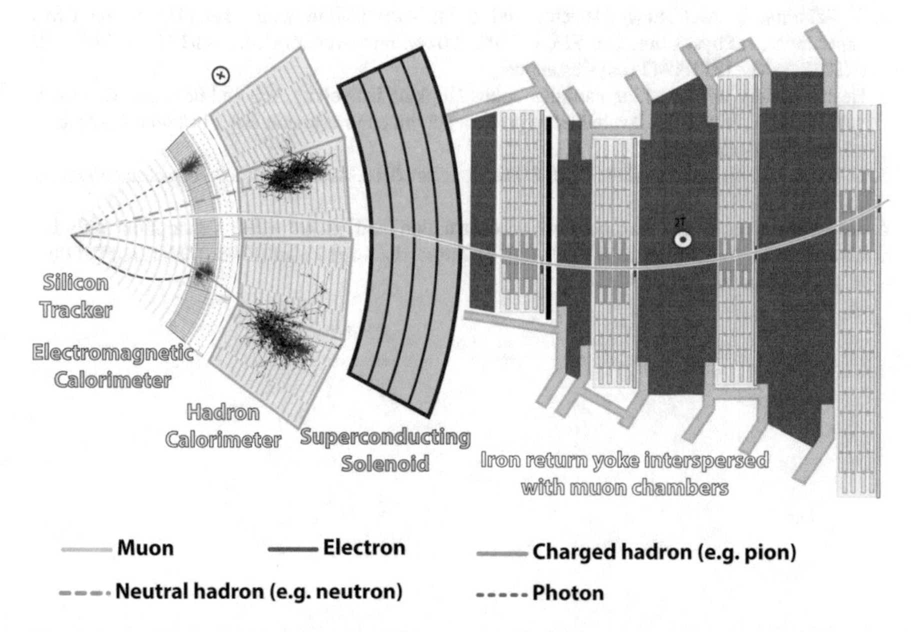

Fig. 2.4 A cross sectional view of the CMS detector, showing where each type of particle leaves its tracks or deposits its energy. This picture is from [5]

deposits in ECAL are photons. The algorithm then reconstructs hadronic jets, using the tracker and HCAL for hadronic particles and ECAL for photons produced by the jet.

Vertexing determines which particles are produced in which collision, as well as which particles are produced from secondary decays of long-lived particles. In addition, when an electron, muon, or photon is produced from the decay of a hadron in a jet, the close proximity between the particle and the jet can be used to determine that, include that particle's momentum in jet calculations, and exclude it from independent consideration.

Jets matching certain criteria, such as a secondary vertex matching the lifetime of B mesons, are tagged as b-jets, which are likely to have come from bottom quarks. Similar considerations are used to tag jets as being produced from τ lepton decays.

The events originally provided by the detector are saved in "raw" format. From there, they are processed into more useful data formats that contain electrons, muons, photons, jets, and MET, which can be used directly in analysis.

References

1. CERN, Aerial views: CERN (2014). Available: https://cern60.web.cern.ch/en/exhibitions/cern (visited on 06 May 2020)
2. CMS Collaboration, Public CMS luminosity information (2019). Available: https://twiki.cern.ch/twiki/bin/view/CMSPublic/LumiPublicResults?rev=162 (visited on 09 Feb 2019)
3. T. Sakuma, T. McCauley, Detector and event visualization with SketchUp at the CMS experiment. J. Phys. Conf. Ser. **513**, 022032 (2014). https://doi.org/10.1088/1742-6596/513/2/022032, arXiv:1311.4942 [physics.ins-det]
4. Heshy Roskes '14: smashing particles. Johns Hopkins University Arts and Sciences Magazine, vol. 11, no. 1, Nov 2013. Available: https://krieger.jhu.edu/magazine/2013/11/06/heshy-roskes-14-smashing-particles/
5. D. Barney, CMS detector slice. CMS Collection, Jan 2016. Available: https://cds.cern.ch/record/2120661
6. A.M. Sirunyan et al., Particle-flow reconstruction and global event description with the CMS detector. J. Instrum. **12**, P10003 (2017). https://doi.org/10.1088/1748-0221/12/10/P10003, arXiv:1706.04965 [physics.ins-det]

Chapter 3
Alignment and Calibration of the CMS Tracker

This chapter describes the calibration of the CMS tracker, and in particular its alignment [1–3]—the position, orientation, and shape of each of its modules as recorded in the CMS reconstruction software. The alignment is a crucial ingredient in translating the information provided directly from the tracker readout, the position of each hit relative to the module, to the information we actually want, the hit's location in 3D space.

For the pixel modules, the precision of a hit measurement is typically a few microns. Therefore, we need to know the module's location to similar precision. This is much smaller than errors that can be introduced when building the detector. In addition, the modules tend to move over time, in particular when the magnetic field is turned on or off or when the temperature changes. Therefore, a regular data-based *alignment* is needed to maintain the precision of the detector. Some changes only cause the large mechanical structures to move; for example, the two halves of BPIX and two halves of each endcap of FPIX are especially sensitive to magnetic field changes. In those cases, only those structures may need to be aligned. The detector is designed in a hierarchy of structures, shown in Fig. 3.1, any of which can be aligned while keeping the relative positions of its components fixed.

Tracker alignment is not like a tire alignment. A tire alignment involves the mechanic jacking up the car and physically moving its wheels to their correct positions and orientations. By contrast, when we align the tracker, we do not go down to the CMS cavern and move any modules, because whether the module positions exactly match the design specifications is not the point. What matters is that the module positions assumed in the track reconstruction match the actual positions of the modules. Alignment can be done months or even years after data taking, and the data can be reconstructed with the new alignment.

On the other hand, it is important that the detector not be *too* misaligned at the time of data taking, because this can affect the track information that goes into the high level trigger. While events that pass the trigger can always be rereconstructed later, events that fail the trigger are lost forever. This happened at the beginning of

J. Roskes, *A Boson Learned from its Context, and a Boson Learned from its End*, Springer Theses, https://doi.org/10.1007/978-3-030-58011-7_3

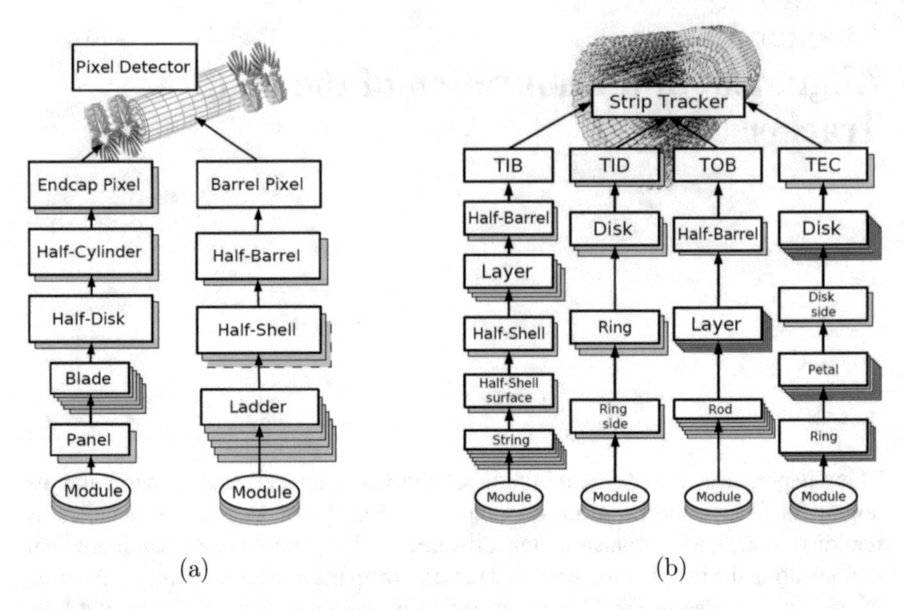

Fig. 3.1 Hierarchies of the pixel (**a**) and strip (**b**) components of the tracker. Any of these levels of the hierarchy can be aligned [1]

the 2015 run, after the detector had been shut down for 2 years. The first cosmic ray data revealed that one side of FPIX was several millimeters away from its assumed position, as shown later in Fig. 3.17a, and as a result the rate of tracks in that side of FPIX was around 50% of the rate in the other side. Many of the events with lost tracks were triggered due to information in the muon systems, but the tracks were not reconstructed by the software because the misalignment was severe enough to ruin the pattern recognition responsible for track finding. those events could be recovered by first performing a rough alignment with the tracks that were not rejected and then rereconstructing the data using the new geometry, matching the individual hits to form a track. Any events that did *not* pass the muon trigger were just gone. Such a large error would probably not happen during the more crucial collision data taking, but a severe enough miscalibration could result in the loss of important events.

The alignment procedure can be done at a hierarchical level: it is possible, for instance, to align large structures while keeping the individual modules attached to those structures fixed. Typically, while CMS is running, an automatic procedure aligns the six pixel structures: two half barrels of BPIX and two half cylinders on each side of FPIX. This procedure, with only 36 degrees of freedom, is simple enough to run without human input. Every few weeks, an alignment of the pixel modules is performed manually. At the end of the year, with the increased statistics of the full year's data, a new alignment is derived covering the entire year. The strips, which are known to be fairly stable and where small movements have less of an impact on track resolution, are typically only aligned at the beginning of the run period and during this full-year alignment.

3.1 Principles of Detector Alignment

The alignment is performed using the data collected. A simple example is shown in Fig. 3.2. When a track is reconstructed with a misaligned geometry, the result is as shown on the right side of the illustration. The expected positions of the hits, calculated from the track's path and shown in red, and the measured hits, shown in green, no longer match. This indicates that the detector is misaligned: its assumed position is wrong.

A real alignment involves many more modules, up to all 17004 (since 2017; the number of modules was previously slightly smaller) for the most comprehensive cases. The assumed positions of any of these modules could be wrong, and we need to fit for all of their positions and rotations in three dimensions. For each module, a local coordinate system is used, as shown in Fig. 3.3. The w axis is perpendicular to the module, and the u and v axes are within it, with the u axis in the more sensitive

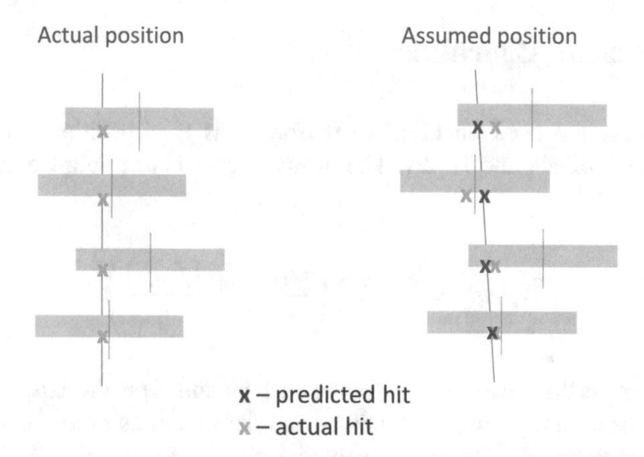

Fig. 3.2 A simplified illustration of how alignment works. The left side shows the actual position of the detector at the time of data taking, with a blue track, taken with the magnetic field turned off, that leaves four hits shown in green. The right side shows the assumed detector position, with the second module from the top assumed to be slightly to the left of its actual position, with the same measured hits in green and the expected hits in red

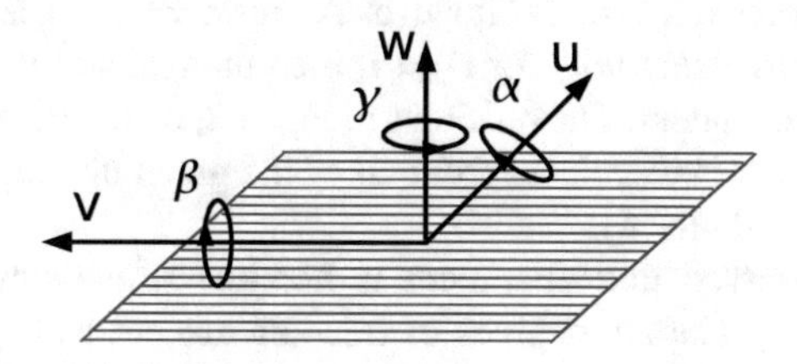

Fig. 3.3 Illustration [2] of the module local coordinates u, v, w and the corresponding rotations α, β, γ for a module. Three other alignment parameters, not shown here, define the curvature of the module

direction of measurement. The angles α, β, and γ describe rotations around u, v, w respectively.

Millions of tracks are used simultaneously for a more precise measurement of all of the module positions. The magnetic field is turned on and the tracks have curvature. Unlike the exact hit positions shown in the picture, the hit measurements have an intrinsic uncertainty, as do the expected hit positions, which are calculated from the other hits on the track. Additionally, the track parameters themselves are affected by the alignment, so a procedure is needed to deal with those correlations as well.

The alignment parameterization in CMS also includes three degrees of freedom for the deformation of the modules, expanding the curved shape of the modules up to quadratic terms: $\Delta w = s_u u^2 + s_v v^2 + s_{uv} uv$. In general, these deformations are determined by the mechanical stresses on the modules and tend not to change over time, so they are aligned infrequently.

3.2 Alignment Algorithms

Two algorithms are used on CMS to perform this minimization and determine alignments: MillePede and HipPy. The ultimate goal is to minimize the objective χ^2 function:

$$\chi^2(\vec{p}, \vec{q}) = \sum_j^{\text{tracks}} \sum_i^{\text{hits}} \left(\frac{m_{ij} - f_{ij}\left(\vec{p}, \vec{q}_j\right)}{\sigma_{ij}} \right)^2 \tag{3.1}$$

where $m_{ij} \pm \sigma_{ij}$ is the measured position of each hit and f_{ij} is the expected position, which depends on the positions, rotations, and deformations of the modules $\vec{p}$ and the track parameters $\vec{q}_j$. (In the case of the pixels, $\vec{f}_{ij}$ and $\vec{m}_{ij} \pm \vec{\sigma}_{ij}$ are two-dimensional vectors with components in the u and the v direction.) We minimize this χ^2 with respect to $\vec{p}$ and $\vec{q}_j$, with the primary goal being $\vec{p}$. Both algorithms start by linearizing f_{ij}, and any nonlinear parts are handled by running iterations. The quantity in the numerator of Eq. (3.1) is the difference between the measured hit and the reconstructed hit and is known as the *residual*. If, as is typically the case, f_{ij} is calculated only using $m_{i'j}$ for $i' \neq i$, then the residual is *unbiased*, because m_{ij} and f_{ij} are independent. The residuals in Fig. 3.2, on the other hand, are biased, because a single track is calculated from all of the green hits m_{ij}, and that track is used to predict the red hits f_{ij}.

It should be stressed that alignment is not just a mathematical problem of minimizing Eq. (3.1). Certain degrees of freedom are not well-constrained by the tracks. Those degrees of freedom are known as weak modes, and some of them will be described in Sect. 3.4. While in theory alignment should never make the situation worse, in practice there can be biases in track reconstruction that lead to false shifts in position. Sometimes this can be useful, as alignment can smooth over effective shifts and recover degraded performance, but in other situations, particularly when the bias causes movement along a degree of freedom that is weakly constrained, this can make the bias in data worse. An example of this will be shown in Sect. 3.4.1.1, where an unknown bias causes a tension between positively and negatively charged collision tracks, with the result that the alignment gets significantly worse in an attempt to compromise between them. Extra constraints need to be added into the procedure to prevent this.

3.2.1 MillePede

The MillePede algorithm [4, 5] does a simultaneous fit for p_i as well as q_j, automatically resolving most of the correlations between track parameters and alignment parameters. The size of the linearized χ^2 matrix is the number of module parameters, $17004 \times 6 \approx 10^5$ for a full scale alignment, plus the number of track parameters, which could be 10^7. However, most of the matrix's entries are 0: the parameters of one track only have direct correlations with the modules hit by that track. This fact allows MillePede to reduce the matrix to a more reasonable $10^5 \times 10^5$ matrix.

In order to improve the computation speed, the MillePede fit runs outside the CMS software package using independent Fortran code. The track propagation model is similar, but not identical, to the standard CMS model. This is both a strength and a weakness of the MillePede approach: the track propagation runs faster and provides an independent cross check of CMS's model, but can also introduce small inconsistencies with the final track reconstruction. In practice, large enough inconsistencies would lead to an incorrect alignment and would be caught during the validation procedure.

3.2.2 HipPy

The HipPy algorithm runs iteratively. In each iteration, it runs over the tracks, using as input the alignment derived in the previous iteration. Subsequently, it aligns each module individually, inverting a simple 6×6 (or 9×9, when the sensor curvature is also aligned) matrix for each module. In practice, when only small, random movements are involved, ten iterations are usually more than enough to deal with correlations between modules.

Figure 3.4 shows the capabilities of the HipPy algorithm. It can take input both from tracks ("AlCaReco") and from other sources of constraints, such as the optical survey or laser alignment system. The optical survey was in active use during Run 1 of the LHC. Although the laser system was decommissioned during Run 2, the functionality can be used to constrain degrees of freedom that tend to move in unphysical directions, such as the ones described in Sect. 3.4.1.1. Reading these

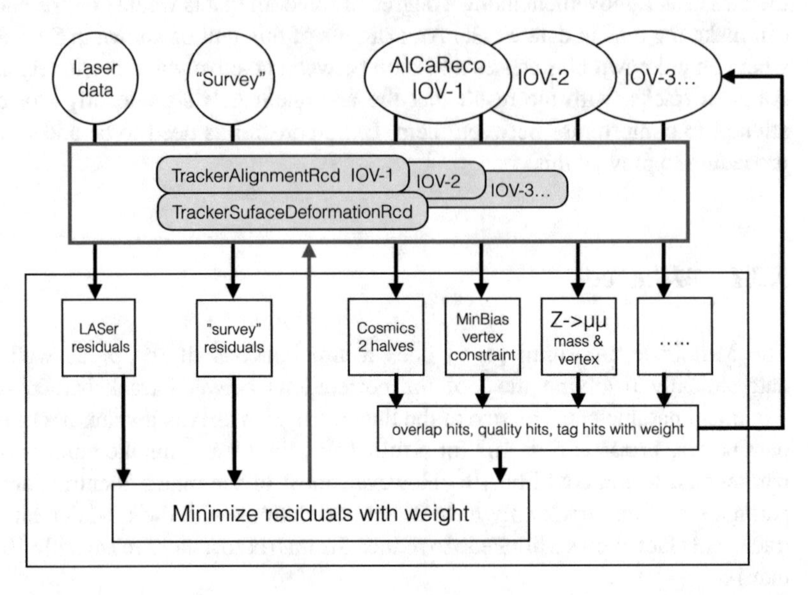

Fig. 3.4 Diagram describing the HipPy alignment procedure

sources of information, and processing them through the initial alignment for the iteration, HipPy calculates residuals, as shown in Fig. 3.2. Because HipPy's track reconstruction has access to the full CMS software, it has the capability to use any type of constraint defined in that software. Certain types of tracks and hits provide additional information—for example, the two tracks $Z \to \mu\mu$ decays can be constrained to come from the same vertex and to have a dimuon mass around 91.2 GeV, and cosmic rays provide a unique topology of tracks that can constrain degrees of freedom not covered well by collisions. These tracks can be weighted higher in the fit in order to best use this information.

In the end, the χ^2 is minimized and the new output is created, which can include the modules' positions and rotations ("TrackerAlignmentRcd") and their curvatures ("TrackerSurfaceDeformationRcd"). Then, the next iteration is run, starting from the output of the previous iteration.

HipPy can also handle multi-IOV alignment, which is necessary when part of the detector shifts at certain points in time and we need to find a separate alignment for each time period, known as an "interval of validity" or IOV. The simplest way to handle this movement would be to simply derive a completely separate alignment for each IOV. However, if we assume that some degrees of freedom remained fixed in all IOVs, we can gain information by using *all* tracks to measure those degrees of freedom, while the other degrees of freedom are aligned separately for each IOV. In HipPy, this is handled by first aligning the components that frequently move, independently for each IOV. Subsequently, another alignment starts from the output of the first step and moves the individual components in a correlated way across all IOVs.

Typically, the large components of the pixel detector move most frequently and the strips move less often. An example alignment procedure might start by aligning the large pixel structures separately in each of 15 IOVs. After those outputs are collected, the pixel *modules* would be aligned within those structures for each of 5 IOVs, each of which spans 3 of the original 15 IOVs. In IOVs 1, 2, and 3, the large structures would be in 3 different places, but the relative positions and rotations of the modules within those structures would be common. Finally, the strip modules, which are known not to move frequently, would be aligned in a single alignment covering all 15 IOVs. The curvatures of all of the modules, which also do not change significantly with time, would also be aligned in the last step.

In dealing with a real systematic global movement, HipPy is slower to converge than MillePede, because the correlations between modules have to be solved through iteration. On the other hand, HipPy is more resilient than MillePede to *false* correlated movements.

3.3 Validation Procedures

Several validations are used to check the effect of alignments and determine whether a particular alignment performs well. A validation is essentially a projection of the alignment performance onto a particularly interesting degree of freedom. The quantities we choose to plot typically have a known value under perfectly aligned conditions. For example, a histogram of residuals is expected to peak at 0 with some width. The difference in parameters between two halves of a cosmic ray track is also expected to be 0 on average. The mass of a reconstructed Z boson should be around 91.2 GeV. By detecting deviations from these expected values, especially deviations as a function of the track location or direction, we search for biases.

This section contains a description of several of the validation procedures used in alignment. Example plots can be found in Sects. 3.4 and 3.5.

3.3.1 Overlaps

The overlap validation monitors the alignment by using hits from tracks passing through regions where modules overlap within a layer of the tracker. In this method, the difference in residual values for the two measurements in the overlapping modules is calculated. Unexpected deviations between the reconstructed hits and the predicted positions can indicate a misalignment. This is characterized by a non-zero mean of the residuals. This method is particularly powerful because the distance between the overlapping modules from the same layer is relatively small, and therefore there is a relatively small uncertainty in track propagation through space between the modules. The double difference in estimated and measured hit positions becomes very sensitive to systematic deformations.

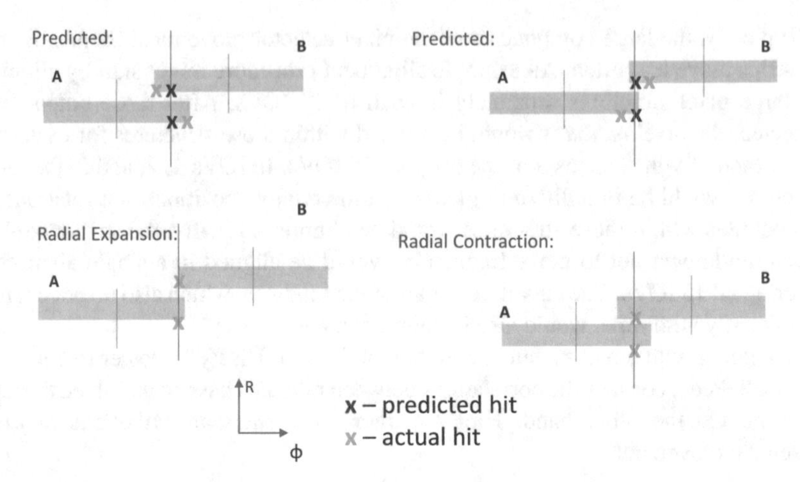

Fig. 3.5 Illustration of the effect on overlaps between modules when they move relative to each other in the plane of the overlap. The top pictures show the module positions in the assumed geometry, with corresponding predicted hits in red. The bottom pictures show the actual positions of those modules, with the reconstructed hits in green. When projected onto the assumed geometry, the two hits are inconsistent, and the residuals have opposite signs between the two modules

Two effects in the overlap validation contribute to detecting misalignments. These effects are illustrated here for the radial misalignment, which will be discussed in detail in Sect. 3.4.5. The first effect, shown in Fig. 3.5, detects movement of the modules within the module plane, which is a second-order effect for the radial misalignment. This leads to a positive shift in residual mean for expansion and a negative shift for contraction, as shown in Fig. 3.5.

A second effect detects common movement of both modules perpendicular to themselves. The sensitivity to this movement comes from the fact that two nearby hits provide a precise measurement of the track angle, and the precision on this measurement, combined with knowledge of the track's momentum from the rest of the hits, can detect the change in track angle resulting from small perpendicular movements. This effect has an opposite effect for positively and negatively charged tracks because they curve in opposite directions in the magnetic field. Therefore, it does not affect the average overlap residual, but might be usable in future studies by isolating tracks of a particular charge. This effect is shown in Fig. 3.6.

3.3.2 *Cosmic Track Splitting*

Cosmic track splitting monitors the alignment of the tracker by independently reconstructing the upper and lower portions of cosmic ray tracks that go through the tracker. It then compares the parameters describing the two paths to see if they match up. This method is also powerful because we know that the two halves of a

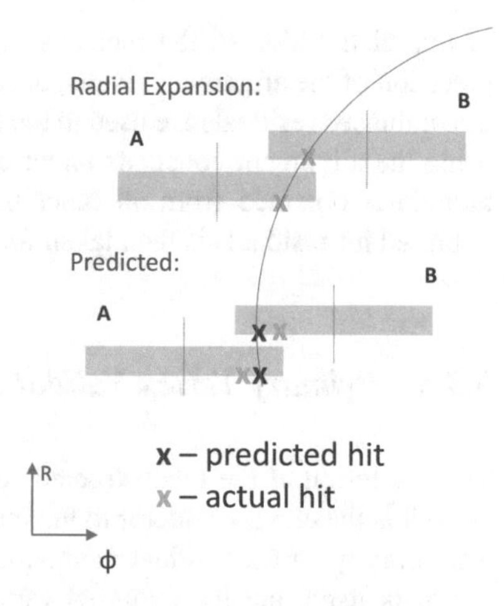

Fig. 3.6 Illustration of the effect on overlaps between modules when they move perpendicular to their plane. The bottom illustration shows the predicted position of the modules with predicted hits in red, while the top illustration shows their actual position with reconstructed hits in green. The radial expansion causes the modules to move along the track and, because of the curvature, to measure a different track angle

given cosmic track should have the same parameters at the origin, while each half of a track mimics a regular collision track originating from the interaction point. Systematic differences between the track halves can indicate a misalignment.

3.3.3 Z → $\mu\mu$ Decays

The Z → $\mu\mu$ validation uses a sample of Z → $\mu\mu$ events and looks for biases by reconstructing the mass of the muon pair. Each event, with its reconstructed mass, is placed into a bin depending on η and ϕ of the muons. The mass distribution of each bin is then fit with a Gaussian, and the mean of this Gaussian is recorded as the reconstructed mass in that bin. The bins are then used to construct profiles of the mass as a function of η or ϕ. Misalignment in the tracker may be detected if the mean reconstructed mass strays from the expected value of 91.2 GeV, either uniformly or as a function of η and ϕ.

3.3.4 Distributions of the Medians of the Residuals (DMRs)

The distribution of the medians of the residuals (DMR) is a powerful tool to assess the statistical precision of alignment. While residuals themselves exhibit natural statistical fluctuations, the mean of those residuals should be zero in the limit of infinite statistics if there is no bias in the alignment and calibration of the detector. However, in order to reduce sensitivity to the tails of the distributions, the median is a better quantity to monitor compared to the mean. With a large enough number of tracks N passing through each module, the median distribution should be centered at zero, and its width should scale as $1/\sqrt{N}$. With a large enough N, the width of this distribution of the medians of residuals (DMR) is a measure of the local precision of the alignment results; deviations of the mean from zero indicate biases. The unbiased residuals are used in the DMR calculations, when each track is refitted using the alignment constants under consideration, and the hit prediction for each module is obtained from all other track hits. The median of the distribution of unbiased hit residuals is then taken for each module and histogrammed.

3.3.5 Primary Vertex Validation

The resolution of the reconstructed vertex position is driven by the pixel detector, since it is the closest detector to the interaction point and has the best hit resolution. The primary vertex residual method is based on the study of the distance between the track itself and the unbiased vertex, which is reconstructed without the track under scrutiny.

The distributions of the unbiased track-vertex residuals in the transverse plane, d_{xy} and in the longitudinal direction, d_z, are studied in bins of track azimuth ϕ and pseudo-rapidity η. Random misalignments of the modules affect only the resolution of the unbiased track-vertex residual, increasing the width of the distributions, but without biasing their mean. Systematic movements of the modules will bias the distributions in a way that depends on the nature and size of the misalignment and of the selected tracks.

3.4 Systematic Misalignments

This section will discuss studies designed to detect systematic misalignments of the tracker, where all modules move in a correlated way. Two basic categories of systematic misalignments arise in alignment:

1. Weak modes are particular degrees of freedom that are difficult to detect using the standard alignment procedures. The most obvious, but not very interesting, example is a global movement of the whole detector: if we reconstructed tracks under the assumption that the entire CMS was transported to the moon, the shape and quality of all tracks would be unchanged. In this section, we study some more interesting cases. For example, a uniform radial expansion of the tracker by a factor $1 + \epsilon$ preserves the shape of tracks as helixes, but introduces biases in the track curvature and hence in the momentum.
2. Biases can also arise due to tension between conflicting constraints used in the alignment procedure. For example, sometimes alignment using cosmic rays and alignment using collisions will find two slightly different optimal positions, and in a real alignment, which uses both types of tracks, the algorithm tries to compromise between them. When this happens, it indicates a bias in

the procedure to reconstruct the tracks, before alignment enters the picture. The information provided by alignment can be used to improve the tracking procedure and, in the meantime, to find the best alignment to use for practical purposes given the non-optimal tracks. Applying weights to different kinds of tracks is a useful strategy in this case, because we can weight each track topology based on the confidence we have in the information provided by that topology.

The goal of this study was to identify systematic misalignments in CMS tracker geometry using various validation tools. The misalignments examined were first order misalignments of $\Delta\phi$, Δr and Δz as functions of z, r and ϕ. Each misalignment was characterized by some ϵ. Systematic misalignments were generated on the ideal geometry using Monte Carlo. For each type of systematic misalignment, four different misalignments were generated using different values of ϵ. The effect of each of the four misalignments was then found in some validation plot for each different systematic misalignment, and a fit was applied to determine the relationship between ϵ and a parameter of that fit.

We determined constraints on these systematic misalignments in the CMS Tracker by comparing the effects of misalignment in simulated Monte Carlo sample and in a representative Run2 data period using both collision and cosmic track data. The two most important validation techniques in this study are the overlap residuals and cosmic tracks split into two halves, have been the focus of this work, following on the original work during the tracker commissioning at the start of Run1 [2]. We have also revisited the systematic z-expansion in the TEC and TOB, following studies from the beginning of Run 2. Data from the 2017D run period, which ran from August 30 to September 20, 2017, was used with one of the intermediate alignments towards the final alignment to be used for reprocessing of the 2017 data. Values of ϵ for each systematic misalignment in this geometry were determined by looking at the parameter identified using the Monte Carlo simulation and using the corresponding fit to identify a characteristic ϵ according to Eq. (3.2).

Let us introduce nine first-order deformations natural for the cylindrical geometry of the CMS tracker and parameterize them with simple models described by a single parameter ϵ. The misalignments in Δz, Δr, $\Delta\phi$ are functions of z, r, ϕ, with an overall scaling given by ϵ. The functional forms used to generate each systematic misalignment are listed in Table 3.1.

For each misalignment, we use the following equation to relate the systematic misalignment plots to ϵ:

$$\text{Quantity from Plot} = a\epsilon + b \tag{3.2}$$

where the quantity from a plot could be, for example, the mean of a distribution or a parameter extracted from a fit. In general, we expect $b = 0$, but we allow this additional degree of freedom in the equation to distinguish alignment issues from other possible effects related to reconstruction.

In describing the ϵ values for misalignments, care must be taken as to the sign. In order to save computing time, Monte Carlo simulations are always done

Table 3.1 Table of the nine basic systematic distortions in the cylindrical system, with the names of each systematic misalignment, the function by which the misalignment is generated, and a validation type sensitive to the misalignment. In the formula for bowing, $z_0 = 271.846$ cm, which is the length of the tracker

	Δz	Δr	$\Delta\phi$
z	**z-Expansion** $\Delta z = \epsilon z$ Overlap	**Bowing** $\Delta r = \epsilon r(z_0^2 - z^2)$ Overlap	**Twist** $\Delta\phi = \epsilon z$ $Z \to \mu\mu$
r	**Telescope** $\Delta z = \epsilon r$ Cosmics	**Radial** $\Delta r = \epsilon r$ Overlap	**Layer rotation** $\Delta\phi = \epsilon r$ Cosmics
ϕ	**Skew** $\Delta z = \epsilon \cos\phi$ Cosmics	**Elliptical** $\Delta r = \epsilon r \cos(2\phi)$ Cosmics	**Sagitta** $\Delta\phi = \epsilon \cos\phi$ Cosmics

using the ideal geometry, and the track reconstruction is done with a possibly misaligned geometry. That is, the "actual" detector position remains fixed to the ideal geometry, and the geometry used in reconstruction changes. When discussing data, the opposite convention is more natural: the geometry used in reconstruction is initially fixed and the actual detector moves.

The equations in Table 3.1 are to the geometry used in reconstruction. Taking the radial misalignment as an example, a value of $\epsilon > 0$ means that the geometry used for reconstruction is expanded in the r direction with respect to the geometry used in data taking. If this happens in data, we call it a radial contraction, because the detector has moved with respect to the expected position. This is the convention used in the text when describing the misalignments in data, as well as in Figs. 3.5 and 3.6 above.

3.4.1 *z*-Expansion

z-expansion (or contraction) is the uniform misalignment of the tracker in the z direction as a function of z. Because the strip barrel modules are blind to the z direction, this misalignment is difficult to detect there. z-expansion in BPIX can be detected using overlaps. We find that a change in ϵ causes a shift in the mean of the overlap validation plot for overlaps in the z direction. The misalignment is an increasing function of ϵ. The effect of the misalignment on the mean of the overlap plot is relatively small.

After fitting the mean of the Overlap Validation distribution in Fig. 3.7 with Eq. (3.2), we have that $a = (-2.83 \pm 0.05) \times 10^4$ μm and $b = (0.58 \pm 0.07)$ μm. We find that in the pixels, the run 2017D ϵ corresponding to z-expansion is $(3 \pm 6) \times 10^{-5}$. In BPIX (at $z = 260$ mm), this corresponds to a contraction of (9 ± 16) μm, consistent with zero.

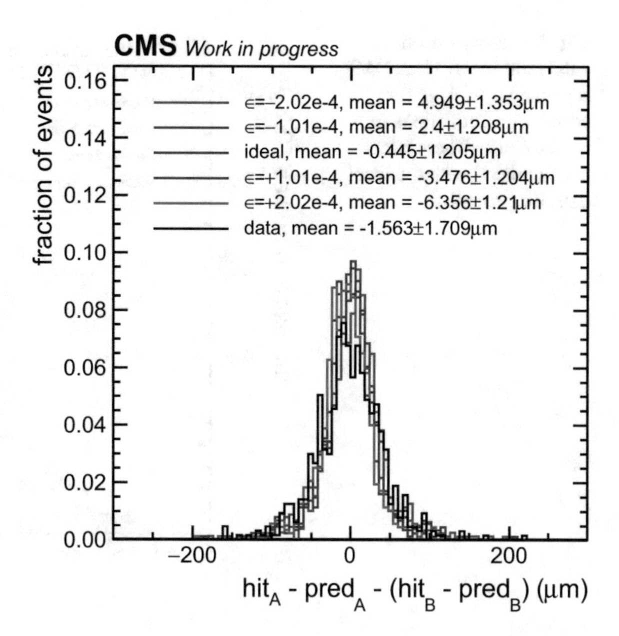

Fig. 3.7 z-expansion validation: distribution of overlaps in the z direction with modules overlapping in the z direction in BPIX for cosmic muon events in Monte Carlo and data. The Monte Carlo events are simulated with the ideal detector geometry and reconstructed using five geometries, corresponding to the z-expansion misalignment with $\epsilon = -2.02 \times 10^{-4}$, -1.01×10^{-4}, 0, 1.01×10^{-4}, and 2.02×10^{-4}

3.4.1.1 z-Expansion in the TEC: DMRs Separated by Charge

In previous alignments, it has been noticed that TEC has experienced some z-expansion. This is caused by a tension between collisions and cosmics, as the collisions appear to show a z-expansion but the cosmics do not. It was also found that in TEC with collision data generated by Monte Carlo, there was a bias between positively and negatively charged tracks, as shown in Fig. 3.8, indicating that there may also be a tension between positive and negative tracks in alignment. A possible explanation could be biased modeling of the track propagation, possibly due to the material model. This suggestion is supported by the fact that the bias is reduced for higher-momentum tracks. Since the same effect appears in both data and MC, it should be possible to track it further with MC simulation. Whatever the source of this effect, it comes from outside alignment, and further study is beyond the scope of this work.

3.4.2 Bowing

Bowing is the misalignment of the tracker in the r direction as a function of z. It is similar to the radial expansion, which will be discussed in Sect. 3.4.5, and differs only by the fact that the bowing effect is a function of z. For small values of ϵ, many millions of events would be needed to measure the z modulation. However, fewer events are needed to *exclude* the presence of either a bowing or a radial misalign-

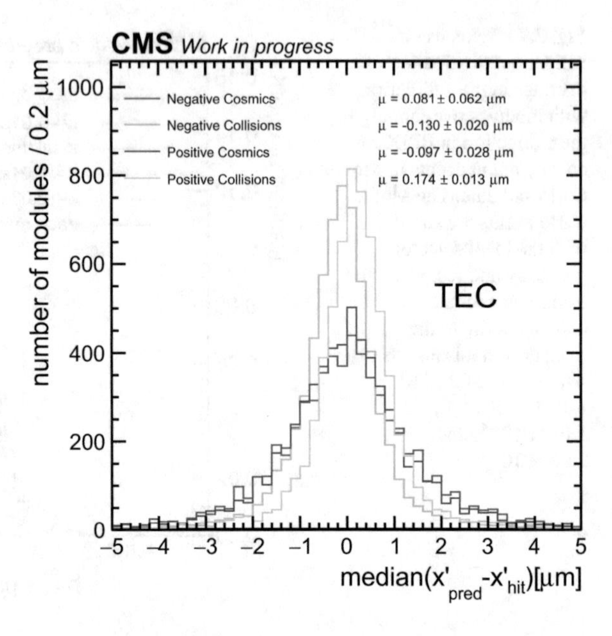

Fig. 3.8 z-expansion validation in the TEC: DMR separated by charge for cosmics and collisions in Monte Carlo. These events are simulated with the ideal detector geometry

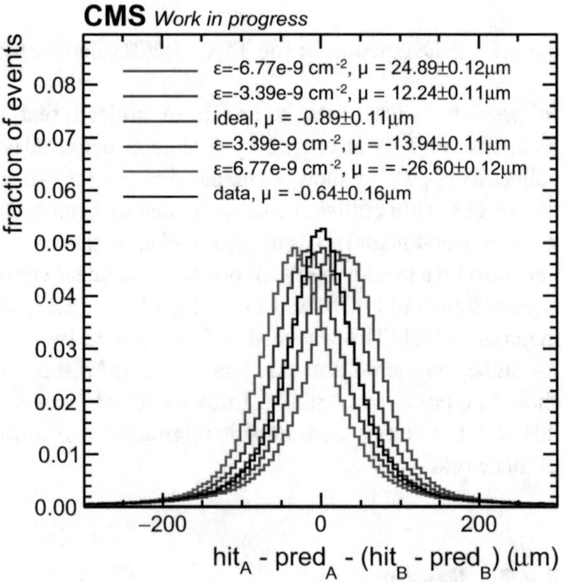

Fig. 3.9 Bowing validation: distribution of overlaps in the ϕ direction with modules overlapping in the ϕ direction in TOB for cosmic muon tracks in Monte Carlo and data. The Monte Carlo events are simulated with the ideal detector geometry and reconstructed using five geometries, corresponding to the Bowing misalignment with $\epsilon = 6.77 \times 10^{-9}\,\text{cm}^2$, $3.385 \times 10^{-9}\,\text{cm}^2$, 0, $-3.385 \times 10^{-9}\,\text{cm}^2$, and $-6.77 \times 10^{-9}\,\text{cm}^2$

ment. There is a clear relationship between ϵ set and the mean value of the overlap distribution, $\mu = \epsilon(-3.816 \pm 0.014) \times 10^9\,\mu\text{m}\,\text{cm}^2 + (-0.86 \pm 0.05)\,\mu\text{m}$. In data, we observe $\mu = (-0.64 \pm 0.16)\,\mu\text{m}$, yielding $\epsilon = (-5.8 \pm 4.5) \times 10^{-11}\,\text{cm}^2$. See Fig. 3.9 for results.

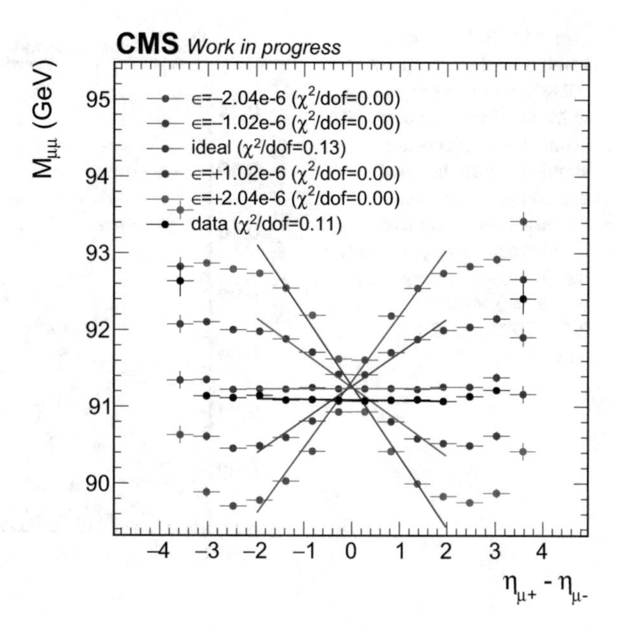

Fig. 3.10 Twist validation: profile of $M_{\mu\mu}$ vs. $\Delta\eta$ for $Z \rightarrow \mu\mu$ events in Monte Carlo and data. The Monte Carlo events are simulated with the ideal detector geometry and reconstructed using five geometries, corresponding to the Twist misalignment with $\epsilon = 2.04 \times 10^{-6}\,\text{cm}^{-1}$, $1.02 \times 10^{-6}\,\text{cm}^{-1}$, 0, $-1.02 \times 10^{-6}\,\text{cm}^{-1}$, and $-2.04 \times 10^{-6}\,\text{cm}^{-1}$

3.4.3 *Twist*

Twist is the misalignment of the tracker in the ϕ direction as a function of z. As such, twist shows up clearly in $Z \rightarrow \mu\mu$ plots, and also in overlap plots. The parameter used is the slope of the $M_{\mu\mu}$ vs. $\Delta\eta$ plot, taken from $\Delta\eta = -2$ to $+2$, as the plot becomes nonlinear for larger $\Delta\eta$. Fitting to the Monte Carlo events, we find that $a = (-4.42 \pm 0.05) \times 10^{-5}$ GeV cm and $b = (-0.018 \pm 0.008)$ GeV. The slope in data was found to be $(-7 \pm 4) \times 10^{-3}$ GeV, corresponding to $\epsilon = (-2.5 \pm 2.2) \times 10^{-8}\,\text{cm}^{-1}$. See Fig. 3.10 for results.

3.4.4 *Telescope*

Telescope is the uniform misalignment of the tracker in the Δz direction as a function of r ($z \rightarrow z + \epsilon r$). This creates concentric rings that are offset in the z-direction, and this misalignment can be visualized by imagining an actual telescope. Because of its z-dependence, Telescope is identified primarily with the track reconstruction of cosmic rays. From fitting Monte Carlo data, we find $a = 3508 \pm 40$ and $b = -0.86 \pm 0.06$. Running this validation on observed data and plugging the mean into Eq. (3.2) yields $\epsilon = (-2.2 \pm 0.5) \times 10^{-5}$. In the pixel detector ($r = 160$ mm) this epsilon corresponds to a maximum relative movement of 3.5 μm, and in the whole tracker ($r = 1100$ mm) it corresponds to a maximum movement of 24 μm. See Fig. 3.11 for results.

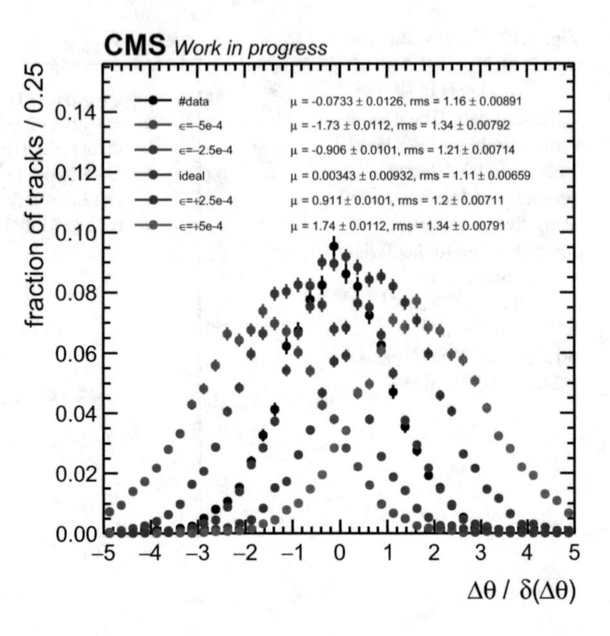

Fig. 3.11 Telescope validation: distribution of $\Delta\theta/\delta(\Delta\theta)$ for cosmic muons in Monte Carlo and data. The Monte Carlo events are simulated with the ideal detector geometry and reconstructed using five geometries, corresponding to the Telescope misalignment with $\epsilon = 5 \times 10^{-4}$, 2.5×10^{-4}, 0, -2.5×10^{-4}, and -5×10^{-4}

3.4.5 *Radial*

Radial is the uniform misalignment of the tracker in the Δr direction as a function of r ($r \rightarrow r + \epsilon r$). Because of the uniform and symmetric nature of this misalignment, it is not easily detected with cosmic track-splitting or $Z \rightarrow \mu\mu$ decays. However, it is easily detected using the Overlap Validation since, in the case of a radial expansion, modules that overlap in the radial direction will move uniformly apart. Therefore, the difference between actual and predicted hit location on two overlapping modules is a good indicator of a radial expansion or contraction. In fact, the linear relationship between the mean of the Overlap Validation plots and the magnitude of the radial misalignment can be used to categorize the presence of radial expansion or contraction in real tracker data.

In TOB, after running the Overlap Validation on Monte Carlo data and fitting the results with Eq. (3.2), we find $a = (-7.461 \pm 0.010) \times 10^4\ \mu\text{m}$ and $b = (-6.023 \pm 0.034)\ \mu\text{m}$. After applying a similar method to tracker data and plugging the mean from the overlap validation into the fit, $\epsilon = (2.23 \pm 0.40) \times 10^{-5}$.

In TIB, we find $a = (-5.035 \pm 0.010) \times 10^4\ \mu\text{m}$, $b = (-3.460 \pm 0.033)\ \mu\text{m}$, and $\epsilon = (-2.82 \pm 0.34) \times 10^{-5}$.

In BPIX, we find $a = (-1.4012 \pm 0.0008) \times 10^4\ \mu\text{m}$, $b = (-1.6450 \pm 0.0030)\ \mu\text{m}$, and $\epsilon = (-9.26 \pm 0.39) \times 10^{-5}$.

Based on the relative epsilon values, we measure a greater radial bias in BPIX than in the other subdetectors. The radius of BPIX is approximately 160 mm, so this corresponds to an overall radial expansion of approximately 15 μm. In TIB (r = 550 mm), we find a *contraction* of 15 μm, and in TOB (r = 1100 mm), we find a *contraction* of 24 μm. See Fig. 3.12 for results.

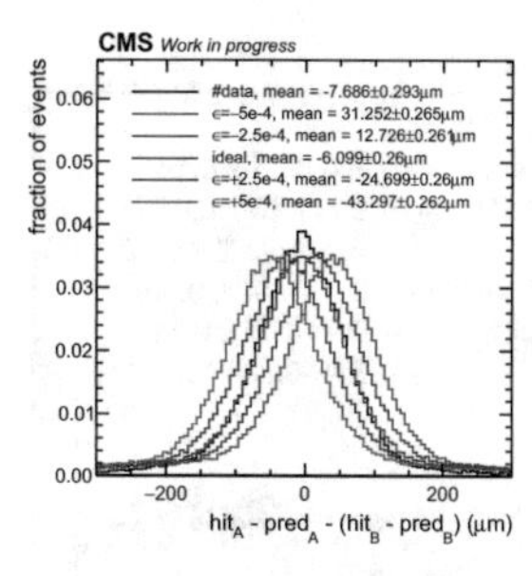

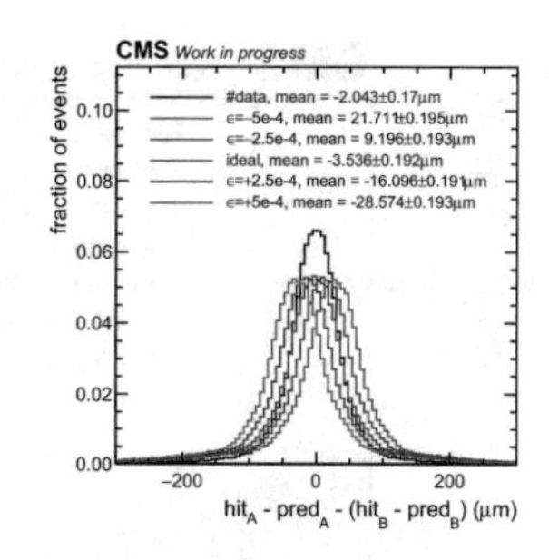

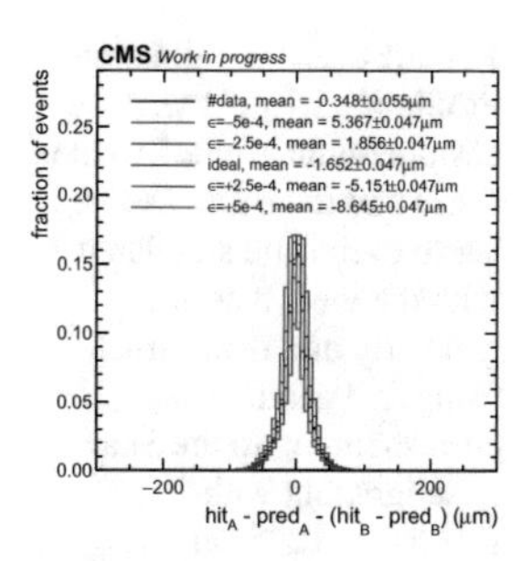

Fig. 3.12 Radial expansion validation: distribution of overlaps in the ϕ direction for modules in the ϕ direction in TOB, TIB, and BPIX for collision events in Monte Carlo and data. The Monte Carlo events are simulated with the ideal detector geometry and reconstructed using five geometries, corresponding to the radial misalignment with $\epsilon = 5 \times 10^{-4}$, 2.5×10^{-4}, 0, -2.5×10^{-4}, and -5×10^{-4}

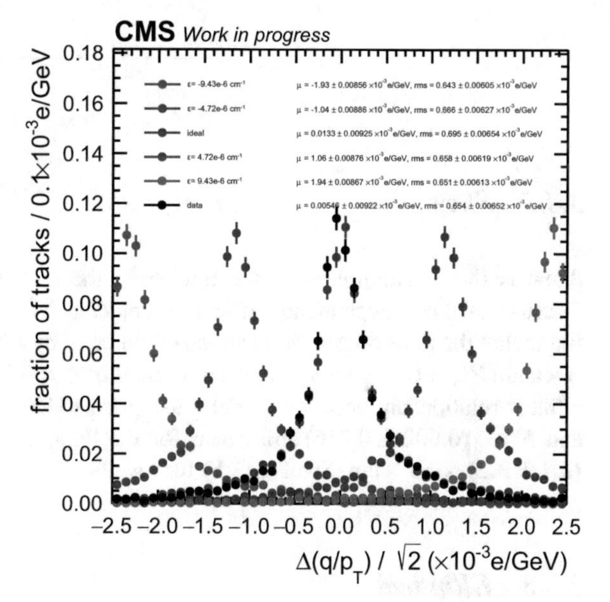

Fig. 3.13 Layer rotation validation: distribution of $\Delta(q/p_t)$ for cosmic muon events in Monte Carlo and data. The Monte Carlo events are simulated with the ideal detector geometry and reconstructed using five geometries, corresponding to the Layer Rotation misalignment with $\epsilon = 9.43 \times 10^{-6}\,\mathrm{cm}^{-1}$, $4.715 \times 10^{-6}\,\mathrm{cm}^{-1}$, 0, $-4.715 \times 10^{-6}\,\mathrm{cm}^{-1}$, and $-9.43 \times 10^{-6}\,\mathrm{cm}^{-1}$

3.4.6 Layer Rotation

Layer rotation is the misalignment of the tracker in the ϕ direction as a function of r. The outer layers twist in one direction, while the inner layers twist in the other direction. The distortion is easily picked up with cosmic track-splitting, as we can see a change in track curvature between the two tracks. As such, we take the mean of a value proportional to the curvature, for each epsilon. We found a linear relationship between μ and ϵ, using Eq. (3.2), with $a = (208.5 \pm 3.9)$ cm e/GeV and $b = (0.9 \pm 2.6) \times 10^{-5}$ e/GeV. For the data, $\mu = (0.005 \pm 0.009)$ e/GeV, so $\epsilon = (-0.2 \pm 1.4) \times 10^{-7}\,\mathrm{cm}^{-1}$ See Fig. 3.13 for results.

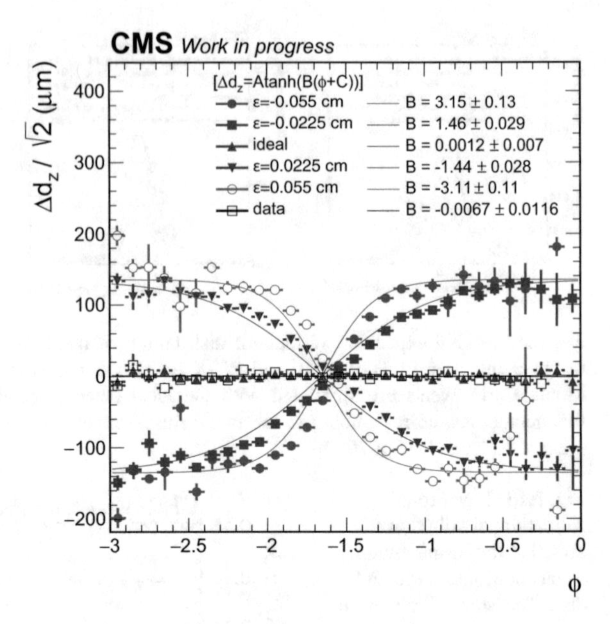

Fig. 3.14 Skew validation: Profile of $\Delta d_z/\sqrt{2}$ vs ϕ for cosmic muon events in Monte Carlo and data. The Monte Carlo events are simulated with the ideal detector geometry and reconstructed using five geometries, corresponding to the Skew misalignment with $\epsilon = 5.5 \times 10^{-2}$ cm, 2.25×10^{-2} cm, 0, -2.25×10^{-2} cm, and -5.5×10^{-2} cm

3.4.7 Skew

Skew is the misalignment of the tracker in the z direction as a function of ϕ. Because of the ϕ dependency, it can be detected with cosmic track splitting. We found that the plots of Δd_z vs. ϕ which could be fit by a hyperbolic tangent function $A \times \tanh(B(\phi+C))$, which can give us ϵ. Setting $A = 134$ and $C = 1.654$, we found a linear relationship between B and ϵ using Eq. (3.2), with $a = (-62.5 \pm 1.9)$ cm and $b = (0.002 \pm 0.016)$ cm. Since for the data, $B = -0.007 \pm 0.012$, $\epsilon = (1.4 \pm 3.2) \times 10^{-4}$ cm. See Fig. 3.14 for results.

3.4.8 Elliptical

Elliptical is the uniform misalignment of the tracker in the Δr direction as a function of $\phi (r \rightarrow r + r\epsilon \cos(2\phi + \delta))$. Because of its ϕ dependency, elliptical is easily detected with cosmic track-splitting. This misalignment is especially clear in the modulation of the difference in the impact parameter Δd_{xy} as a function of the track's angle ϕ. We fit this modulation to a sine function, $\Delta d_{xy} = -A \times \sin(2\phi)$, and find a linear relationship between A and ϵ. Using Eq. (3.2), we find $a = (8.63 \pm 0.11) \times 10^4$ μm and $b = (-0.22 \pm 0.34)$ μm. Using Eq. (3.2), this yields $\epsilon = 2.5 \pm 0.6 \times 10^{-5}$. In the pixel detector ($r = 160$ mm), this ϵ corresponds to a maximum movement of 4 μm, and in the whole tracker ($r = 1100$ mm) it corresponds to a maximum movement of 30 μm. The positive sign of ϵ means that

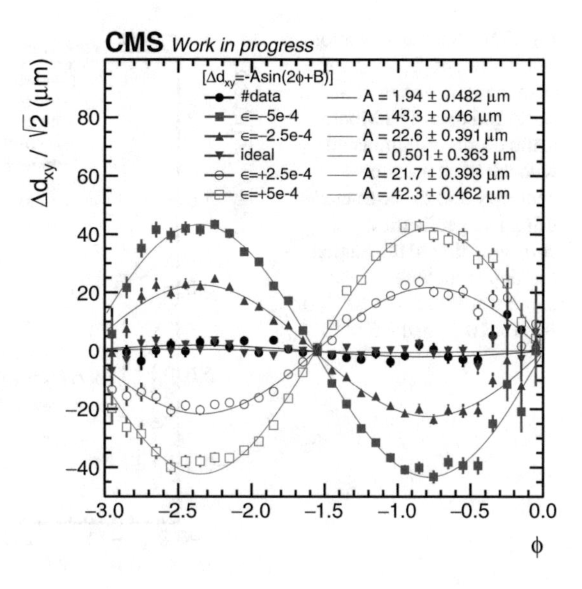

Fig. 3.15 Elliptical validation: profile of $\Delta d_{xy}/\sqrt{2}$ vs. ϕ for cosmic muon events in Monte Carlo and data. The Monte Carlo events are simulated with the ideal detector geometry and reconstructed using five geometries, corresponding to the elliptical misalignment with $\epsilon = 5 \times 10^{-4}$, 2.5×10^{-4}, 0, -2.5×10^{-4}, and -5×10^{-4}

there is a expansion in Δr as a function of ϕ, with the long axis of the resulting oval shape is in the y direction. See Fig. 3.15 for results.

3.4.9 Sagitta

Sagitta is the uniform misalignment of the tracker in the $\Delta\phi$ direction as a function of ϕ. As with the elliptical misalignment, the ϕ dependence in sagitta allows it to be detected with the cosmic track-splitting validation. The effect of the misalignment can be seen in plots with $\Delta\phi$ vs ϕ. Figure 3.16 shows sinusoidal distributions, fit to $-A \times \cos(\phi + B)$, We fit to Eq. (3.2) using the amplitude (A) of the sine wave as the quantity from the plot. Fitting to Monte Carlo, we find that $a = 1199 \pm 5$ and $b = (-2 \pm 2) \times 10^{-3}$. For data, we find that $A = (0.052 \pm 0.009)\,\mu\text{m}$ and thus $\epsilon = (4.5 \pm 0.7) \times 10^{-5}$ mrad. See Fig. 3.16 for results.

3.4.10 Summary

In this section, we have introduced nine first-order deformations of the CMS tracker geometry natural for the cylindrical geometry and parameterized them with the simple models described by a single parameter ϵ. We have determined constraints on these systematic misalignments by examining the effects of misalignment in simulated Monte Carlo sample, and then comparing to collision and cosmic track data from Run2. A characteristic ϵ value, describing the magnitude of a

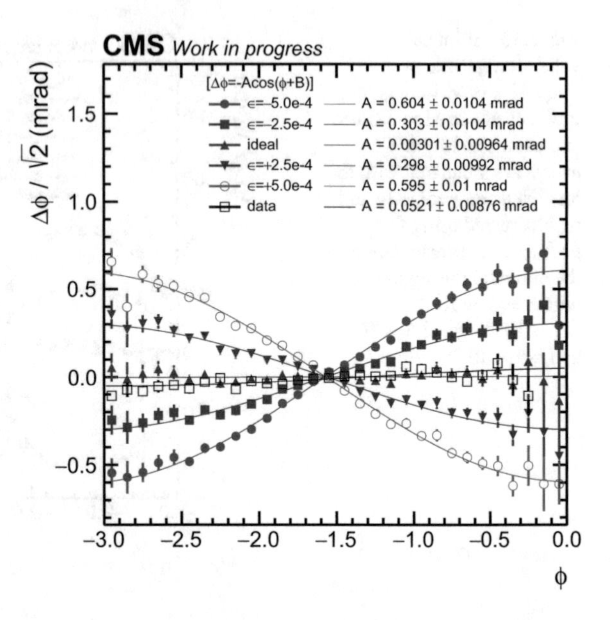

Fig. 3.16 Sagitta validation: distribution of $\Delta\phi$vs. ϕ for cosmic muon events in Monte Carlo and data. The Monte Carlo events are simulated with the ideal detector geometry and reconstructed using five geometries, corresponding to the Sagitta misalignment with $\epsilon = 5 \times 10^{-4}$, 2.5×10^{-4}, 0, -2.5×10^{-4}, and -5×10^{-4}

misalignment in each of the nine scenarios, has been determined for each of the systematic misalignments, along with a corresponding upper limit on the magnitude of ϵ. These results are summarized in Table 3.2. The constraints are presented at 68% CL (1σ). The obtained constraints could be used in physics analyses sensitive to systematic distortions in the tracker geometry to set limits on possible biases.

It may be possible that there is a systematic misalignment present in the tracker that is not represented by any of the nine misalignments studied in this note. However, such a misalignment would likely be a higher order function of z, r or ϕ than the ones used in this study and would likely still have some first order component that would appear in the validation plots described in this note. It would still be useful to examine such systematic misalignments to better characterize the systematic misalignments in the tracker.

One indication of a higher order systematic misalignment would be potential differences between different kinds of tracks or different kinds of plots in estimating the magnitude of the same misalignment. These differences could indicate that the misalignment is not exactly of the form studied or could indicate biases in track reconstruction, unrelated to alignment, similar to the ones seen in Sect. 3.4.1.1.

While we do not pursue this in detail in this study, in the case of radial expansion (Sect. 3.4.5) we obtained three different estimates of ϵ, one each in BPIX, TIB, and TOB, and found different ϵ values, so the misalignment appears to be a higher order function of r. Additionally, because bowing and radial are so similar, we can compare our estimate of ϵ for bowing in TOB, obtained using cosmic rays, and our estimate of ϵ for radial in TOB, obtained using collision tracks. In the center plane of the detector, where bowing's effect is largest, a bowing misalignment with $\epsilon_b = \epsilon$ is equivalent to a radial misalignment with $\epsilon_r = \epsilon_b z_0^2$. Using $z_0 = 271.846$ cm and

Table 3.2 Summary table of ϵ in each misalignment, each misalignment is listed with its corresponding validation type and a maximum amplitude of ϵ (at 68% CL, or 1σ)

	Δz	Δr	$\Delta \phi$
z	**z-Expansion** $\Delta z = \epsilon z$ Overlap $\epsilon = (3 \pm 6) \times 10^{-5}$ $\lvert\epsilon\rvert < 9 \times 10^{5}$	**Bowing** $\Delta r = \epsilon r(z_0^2 - z^2)$ Overlap $\epsilon = (-5.8 \pm 4.5) \times 10^{-11}\,\text{cm}^{-2}$ $\lvert\epsilon\rvert < 1.0 \times 10^{-10}\,\text{cm}^{-2}$	**Twist** $\Delta \phi = \epsilon z$ $Z \rightarrow \mu\mu$ $\epsilon = (-2.5 \pm 2.2) \times 10^{-8}\,\text{cm}^{-1}$ $\lvert\epsilon\rvert < 4.7 \times 10^{-8}\,\text{cm}^{-1}$
r	**Telescope** $\Delta z = \epsilon r$ Cosmics $\epsilon = (2.18 \pm 0.48) \times 10^{-5}$ $\lvert\epsilon\rvert < 2.7 \times 10^{5}$	**Radial** $\Delta r = \epsilon r$ Overlap $\epsilon = (-9.26 \pm 0.39) \times 10^{-5}$ $\lvert\epsilon\rvert < 9.9 \times 10^{5}$	**Layer rotation** $\Delta \phi = \epsilon r$ Cosmics $\epsilon = (-0.2 \pm 1.4) \times 10^{-7}\,\text{cm}^{-1}$ $\lvert\epsilon\rvert < 1.6 \times 10^{-7}\,\text{cm}^{-1}$
ϕ	**Skew** $\Delta z = \epsilon \cos \phi$ Cosmics $\epsilon = (1.4 \pm 3.2) \times 10^{-4}\,\text{cm}$ $\lvert\epsilon\rvert < 4.6 \times 10^{-4}\,\text{cm}$	**Elliptical** $\Delta r = \epsilon r \cos(2\phi)$ Cosmics $\epsilon = (2.5 \pm 0.6) \times 10^{-5}$ $\lvert\epsilon\rvert < 3.1 \times 10^{5}$	**Sagitta** $\Delta \phi = \epsilon \cos \phi$ Cosmics $\epsilon = (4.5 \pm 0.7) \times 10^{-5}$ $\lvert\epsilon\rvert < 5.2 \times 10^{5}$

our measured value for ϵ_b, we find $\epsilon_r = (-4.2 \pm 3.6) \times 10^{-6}$, which is 2σ away from our direct estimate using collision tracks, $\epsilon_r = \epsilon = (2.23 \pm 0.40) \times 10^{-5}$. This may indicate that the misalignment has some other position dependence that affects cosmic rays differently from collisions, though to make a more definite statement it would be necessary to run over more events.

It is interesting to note that four systematic misalignments were found with ϵ inconsistent with zero at very high confidence level: Telescope, Radial, Elliptical, and Sagitta. This may indicate either some time-dependence in the systematic distortions within a given IOV, or more likely some tension between different constraints in the alignment procedure. The observed effects are still small and would not affect most of the physics analyses on CMS, but further investigation of these effects will be a natural continuation of these studies for further refinement of the alignment procedure.

3.5 Performance During Run II of the LHC

This section will cover some of the alignment results throughout Run 2, which ran from 2015–2018. In each year, a selection of plots are shown, so that each type of validation is covered between the 4 years. For comprehensive plots for each year, see [6–9].

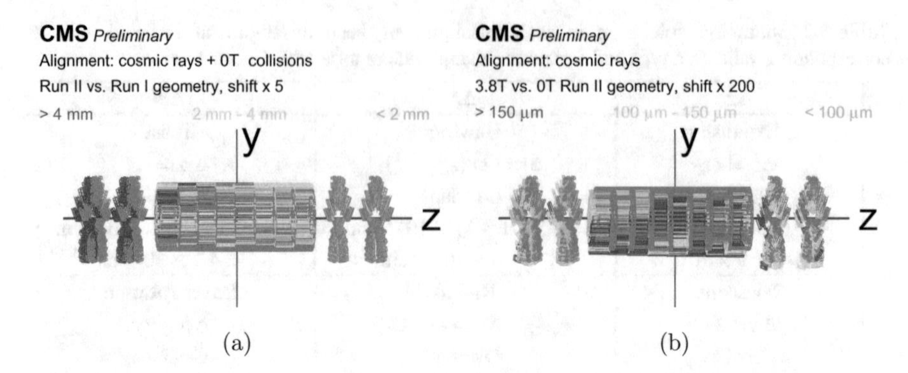

Fig. 3.17 Illustration of the differences in the pixel detector position (**a**) between the end of Run 1 of the LHC and the beginning of Run 2, and (**b**) between the cosmic ray data collection with the magnetic field turned off and with the field turned on. In both plots, the older module positions are shown in gray and the new positions are shown in brighter colors. the colors indicate which modules moved the most, but the color scaling is different between the plots to better illustrate the scale of the movements [6]

3.5.1 2015 Startup

The 2015 run of the LHC saw the first collisions at 13 TeV. It was primarily a preparation run, with only 2.7 fb^{-1} of collisions. During the long shutdown since Run 1, the detector had been opened and the pixel detector was completely removed and replaced, so large movements were expected. Figure 3.17a shows the differences in the pixel detector between the end of Run 1 and the beginning of Run 2. The larger movements are seen in the $-z$ forward pixel detector, which was inserted a few millimeters away from its previous position. The alignment result was the first indication that this had happened. BPIX is mostly yellow in this plot due to a recentering procedure that was performed. Figure 3.17b shows the much smaller movements that resulted from turning on the magnetic field.

3.5.2 2016

The 2016 run produced the first higher luminosity 13 TeV proton collisions. As one of the 3 years of Run 2 designed for precise physics analyses, it was important to maintain the performance throughout the year. Figure 3.18 shows primary vertex validation plots for the 2016 data, comparing the performance before and after the alignment was performed.

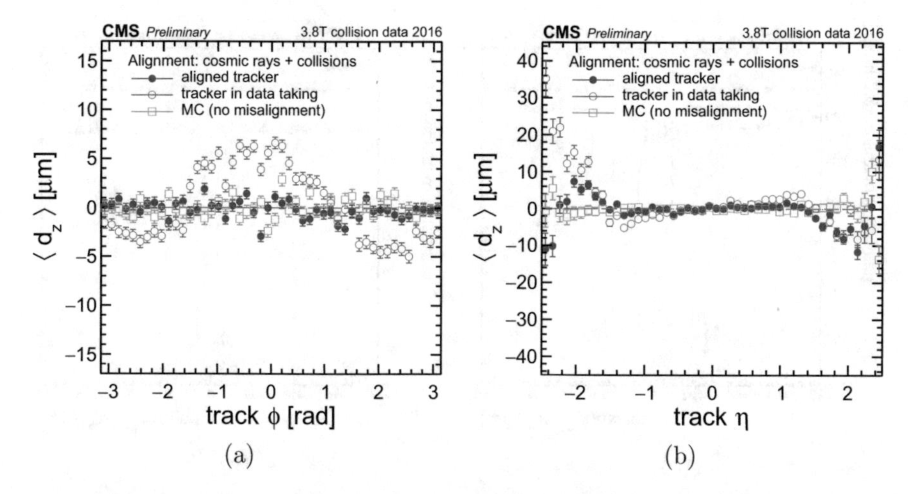

Fig. 3.18 Primary vertex validation plots for 2016 data, comparing the initial alignment used for data taking in red, the alignment used to rereconstruct the data for data analyses in blue, and the Monte Carlo simulation reconstructed under ideal conditions in green. The z distance between the probe track and the vertex is plotted as a function of (**a**) ϕ and (**b**) η of the probe track [7]

3.5.3 2017

For the 2017 run, an entirely new pixel detector was installed. This detector was designed to provide better resolution for tracks by adding an additional BPIX layer closer to the beam pipe and an additional disk on each side of FPIX. At the beginning of 2017, this detector had to be aligned from scratch. The DMR plots in Fig. 3.19 show the improvement resulting from the alignment, first using cosmic rays and then using the first collisions. The plot in Fig. 3.20 shows the effect of the alignment on ϕ modulation of the reconstructed Z boson mass. This kind of modulation is characteristic of a weak mode effect, described in Sect. 3.4, and is fixed by the alignment.

3.5.4 2018

Figure 3.21 shows a comparison of track splitting performance on 2018 cosmic ray data between the alignment at the end of 2017 and at the beginning of 2018.

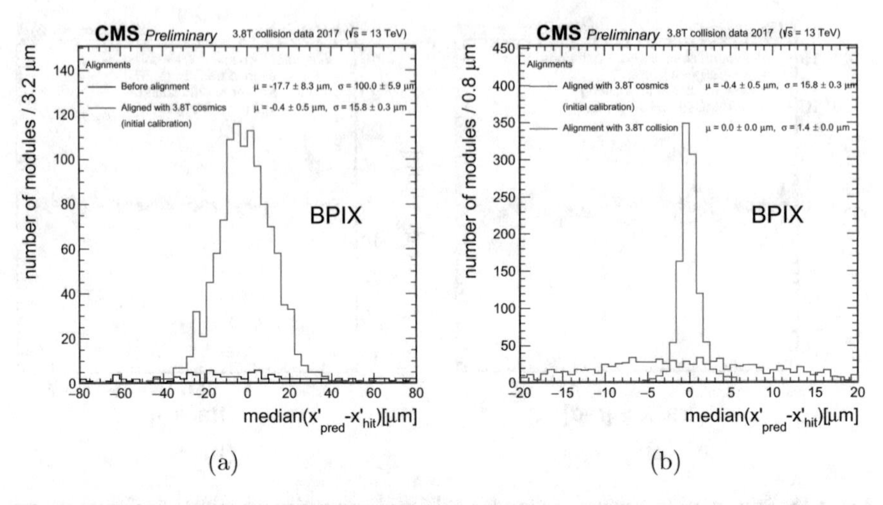

Fig. 3.19 DMR plots for 2017 collision data, comparing three different alignments: the initial geometry used for data taking (black), the first calibration of the detector using cosmic rays (blue), and the updated alignment derived using collision data (red). The alignment with cosmic rays significantly improves the performance, and the alignment with collisions, sensitive to different degrees of freedom that are relevant to the collision tracks used in the validation, brings further improvements [8]

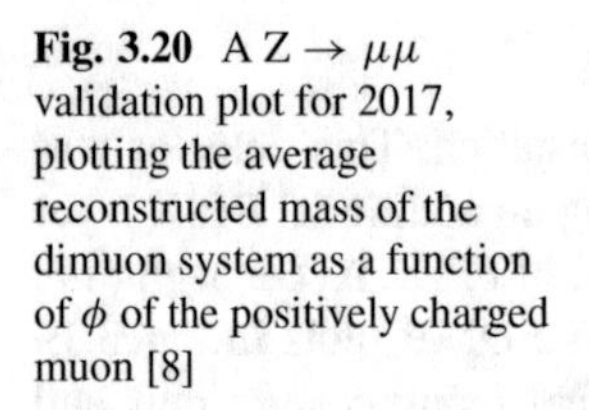
Fig. 3.20 A $Z \to \mu\mu$ validation plot for 2017, plotting the average reconstructed mass of the dimuon system as a function of ϕ of the positively charged muon [8]

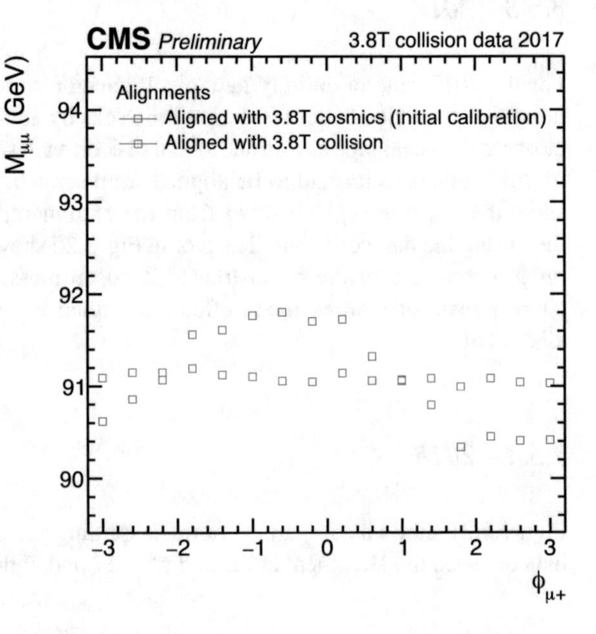

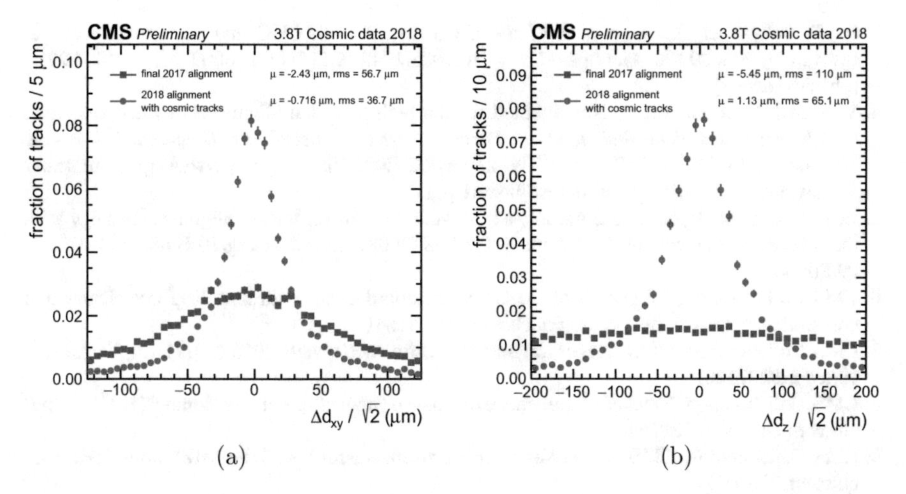

Fig. 3.21 Track splitting plots, showing histograms of Δd_{xy} and Δd_z for the 2018 alignment, reconstructed with the alignment from the end of 2017 and the alignment derived in 2018 [9]

3.6 Conclusions and Outlook

The results shown here are just a small selection of the plots produced by the CMS tracker alignment group over Run 2 of the LHC. The detector conditions changed from what they were in Run 1, first with the increased collision energy in 2015 and subsequently with the new pixel detector in 2017. The conditions will become even more challenging at the High Luminosity LHC (HL-LHC), when the luminosity delivered by the LHC, and hence the number of simultaneous collisions, will increase drastically. The plan is to upgrade the tracker again at that time, including a much more extensive forward pixel detector that can handle the large numbers of particles produced close to the beam line, and the forward degrees of freedom are among the most difficult to align.

Extensive studies are ongoing and will continue throughout Run 3 in order to prepare for these conditions, which will be more challenging than any faced so far. The alignment group and procedures have proven to be flexible and resilient to date, and should be able to incorporate the new developments needed to deliver fast and precise alignments throughout the run period of the HL-LHC.

References

1. W. Adam et al., Alignment of the CMS silicon strip tracker during standalone commissioning. J. Instrum. **4**, T07001 (2009). https://doi.org/10.1088/1748-0221/4/07/T07001. arXiv: 0904.1220 [physics.ins-det]
2. S. Chatrchyan et al., Alignment of the CMS silicon tracker during commissioning with cosmic rays. J. Instrum. **5**, T03009 (2010). https://doi.org/10.1088/1748-0221/5/03/T03009. arXiv: 0910.2505 [physics.ins-det]

3. S. Chatrchyan et al., Alignment of the CMS tracker with LHC and cosmic ray data. J. Instrum. **9**, P06009 (2014). https://doi.org/10.1088/1748-0221/9/06/P06009. arXiv: 1403.2286 [physics.ins-det]
4. V. Blobel, C. Kleinwort, A New method for the high precision alignment of track detectors, in *Advanced Statistical Techniques in Particle Physics. Proceedings, Conference, Durham, UK, March 18-22, 2002* (2002). arXiv: hep-ex/0208021 [hep-ex]. http://www.ippp.dur.ac.uk/Workshops/02/statistics/proceedings//blobel1.pdf
5. G. Flucke, P. Schleper, G. Steinbruck, M. Stoye, CMS silicon tracker alignment strategy with the Millepede II algorithm. J. Instrum. **3**, P09002 (2008). https://doi.org/10.1088/1748-0221/3/09/P09002
6. CMS Collaboration, Alignment of the CMS tracking-detector with first 2015 cosmic-ray and collision data. (2015). http://cds.cern.ch/record/2041841
7. CMS Collaboration, CMS tracker alignment performance results 2016 (2017). http://cds.cern.ch/record/2273267
8. CMS Collaboration, Tracker alignment performance plots after commissioning (2017). http://cds.cern.ch/record/2297526
9. CMS Collaboration, CMS tracker alignment performance results 2018 (2018). https://cds.cern.ch/record/2650977

Chapter 4
Phenomenology of Higgs Boson Interactions

In the SM, once the Higgs boson's mass is measured, all of its other properties are fixed. This chapter will discuss those properties for a Higgs boson with a mass of 125 GeV and parameterize possible deviations that could be produced by BSM effects. This sets the stage for analyses that search for those deviations, which will be described in the next chapter.

4.1 Production Modes

The Higgs boson can be produced through several different processes. The kinematics of the Higgs boson and any associated particles are different for each of these production modes, and for this reason each one provides a unique handle on the Higgs boson's couplings and properties.

The four most common Higgs boson production mechanisms at the LHC are shown in Fig. 4.1. The first, gluon fusion, is expected under the Standard Model to account for 87 % of Higgs boson events in 13 TeV proton-proton collisions [1]. Because gluons are massless, they do not couple directly to the Higgs boson, so gluon fusion proceeds through a quark loop. The primary contribution has a top quark in the loop, although there is also a small contribution from the bottom quark, which affects the p_T spectrum. At leading order, gluon fusion does not produce any other particles together with the Higgs boson. At higher orders, jets can be produced from the initial state gluons through QCD interactions. These jets are typically soft, as is characteristic of jets produced through QCD effects.

The second most common production mechanism is vector boson fusion, or VBF, which accounts for 7 % of Higgs boson events at 13 TeV. In VBF production, two quarks, one from each of the two incoming protons, radiate Z or W bosons, which interact to produce the Higgs boson. The exchange of vector bosons typically involves several hundred GeV of momentum, which is reflected in the large

J. Roskes, *A Boson Learned from its Context, and a Boson Learned from its End*, Springer Theses, https://doi.org/10.1007/978-3-030-58011-7_4

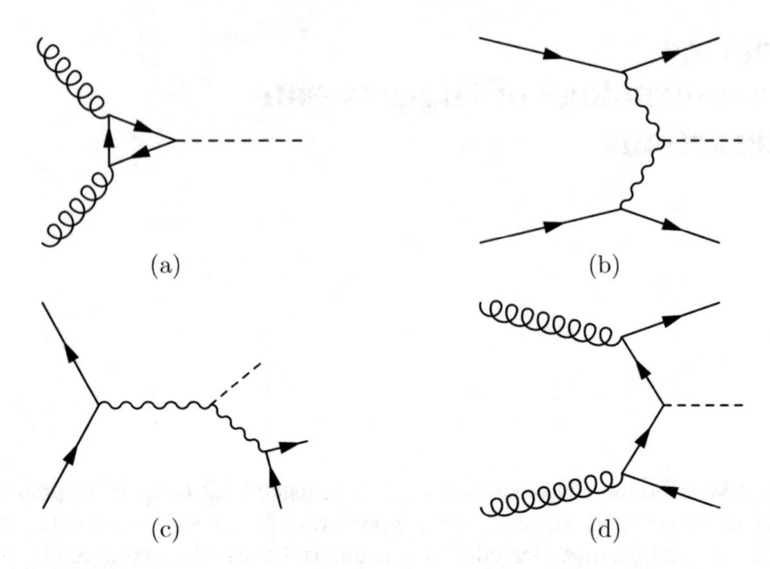

Fig. 4.1 Feynman diagrams for the most common Higgs boson production mechanisms at the LHC: (**a**) gluon fusion, (**b**) vector boson fusion, (**c**) associated VH production, and (**d**) $t\bar{t}H$ or $b\bar{b}H$ production

observed transverse momentum of the two quarks, detected as jets, and the Higgs boson's decay products. The two jets are usually observed traveling in opposite directions.

A further 4 % of Higgs boson events are produced through associated VH production, which again involves the interaction between the Higgs boson and vector bosons. In this case, two quarks from the incoming protons produce an offshell Z or W boson, which then radiates a Higgs boson. As in the case of VBF, additional particles, the decay products of the Z or W boson, are observed together with the Higgs boson. These could be quarks (detected as jets), leptons, or neutrinos (detected as missing transverse energy). The invariant difermion mass of the associated particles is around 80.4 GeV or 91.2 GeV, the masses of the W and Z bosons.

Most of the remaining 2 % of the Higgs boson cross section is comprised of $t\bar{t}H$ and $b\bar{b}H$ production, each of which contributes around 0.9 %. These production modes include two extra top or bottom quarks in the final state, which can be detected by standard methods for identifying top quark decay products or b-jets, respectively. Although these two processes have similar cross sections, $t\bar{t}H$ is easier to detect at the LHC because the top quark's heavier mass gives the associated particles a larger momentum. The b-jets produced in $b\bar{b}H$ are often too soft to detect, and the p_T spectrum of $b\bar{b}H$ events is very similar to that of gluon fusion events.

Other production modes include gg $\rightarrow$ ZH and tqH, which contribute 0.2 % and 0.1 %, respectively, to the total Higgs boson cross section. Because they are so rare, they are not studied here. However, these production modes are particularly

Table 4.1 Summary of the Higgs boson production modes discussed in this section

Production mode	Couplings	Extra particles	Percent contribution at 13 TeV
ggH	Fermion	None or QCD jets	87 %
VBF	Vector boson	Two quarks	7 %
VH	Vector boson	Z or W	4 %
$t\bar{t}H$	Fermion	Two top quarks	0.9 %
$b\bar{b}H$	Fermion	Two bottom quarks	0.9 %
$gg \to ZH$	Both	Z	0.2 %
tqH	Both	Top and light quark	0.1 %

interesting because they can be produced through the Higgs boson's couplings to either vector bosons or fermions, and as such they can be used to measure the interference between these two couplings.

These production modes are summarized in Table 4.1.

Because it has a finite width, the Higgs boson can also be produced offshell, away from its pole mass of 125 GeV. In the offshell region above 200 GeV, the important production modes to consider are ggH, VBF, and VH. $t\bar{t}H$ and $b\bar{b}H$ production become less important at higher mass.

4.2 Decay Modes

The Higgs boson can decay to any light enough massive particle and, through loops, to massless particles as well. The branching fractions for the primary decay modes, for a mass range that includes both 125 GeV and the offshell region, are shown in Fig. 4.2. The Higgs boson decays can be divided into four basic cases, shown in Fig. 4.3:

(a) Simple decays to two fermions. Because the Higgs boson's coupling to other particles is proportional to their mass, its strongest couplings is to the heaviest fermions. The top quark's mass is much higher than that of the Higgs boson, and therefore its decay is kinematically forbidden. Therefore, the most important fermionic decays are to the next heaviest fermions, $b\bar{b}$ or $\tau\tau$.

- The $H \to b\bar{b}$ decay is the dominant Higgs boson decay channel for a mass of 125 GeV, but because this decay only produces b jets, it has large backgrounds from QCD processes and is difficult to isolate. It has been observed by both CMS [2] and ATLAS [3] by looking at leptonic VH production. Because there are extra leptons present in the event, the background for this process is low enough to be manageable.
- The $H \to \tau\tau$ decay has a lower rate and lower background. This channel has also been observed by CMS [4] and ATLAS [5]. The τ leptons may decay either to a lighter lepton and two neutrinos or to one or more mesons and a neutrino. This decay channel will be discussed further in Sect. 5.5.

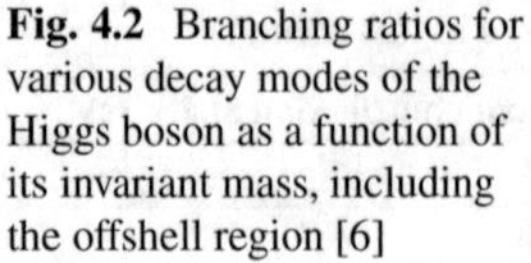

Fig. 4.2 Branching ratios for various decay modes of the Higgs boson as a function of its invariant mass, including the offshell region [6]

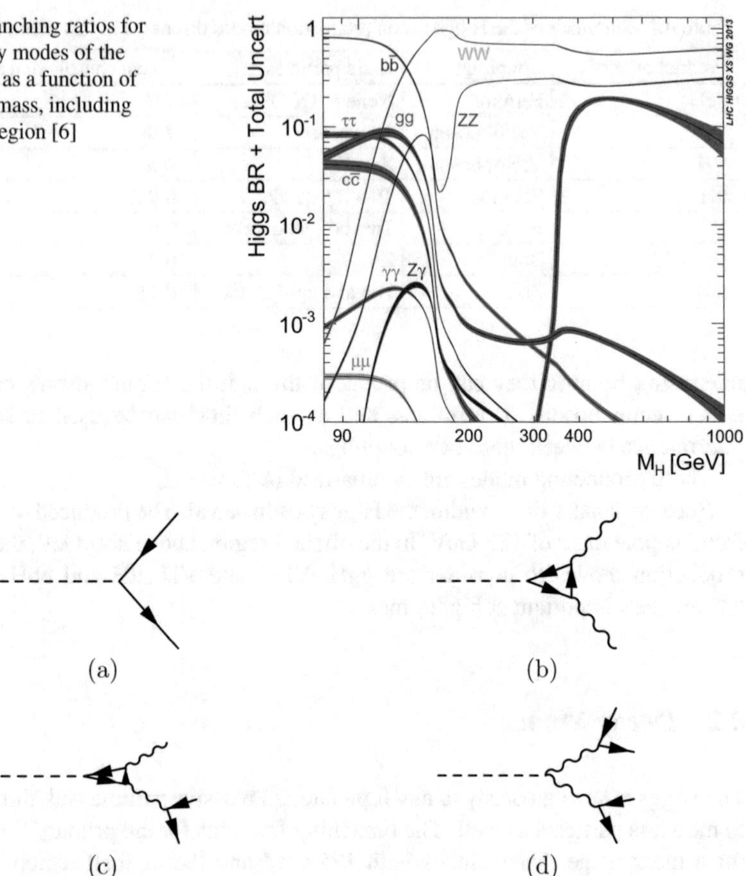

Fig. 4.3 Feynman diagrams for three types of Higgs boson decay modes: (**a**) $H \rightarrow f\bar{f}$ (such as $\tau^+\tau^-$ or $b\bar{b}$), (**b**) $H \rightarrow \gamma\gamma$, (**c**) $H \rightarrow Z\gamma \rightarrow 2f\gamma$, and (**d**) $H \rightarrow VV \rightarrow 4f$ (such as $ZZ \rightarrow 4\ell$)

(b) Decays to two photons. Because photons are massless, the decay proceeds through a top quark or W boson loop. This channel has a lower branching fraction because the loop suppresses the decay, but has the advantage of low background, and was one of the first channels used to discover the Higgs boson in 2012. The Higgs boson decay to two gluons is a similar diagram and also proceeds through a top loop, but because events with two jets are ubiquitous at the LHC, it is an extremely difficult channel to measure experimentally.

(c) Decays to a Z boson and a photon, where the Z boson subsequently decays to two photons. This decay also proceeds through a top quark or W boson loop. The SM rate for the $Z\gamma$ decay is lower than can be measured with the data collected so far.

(d) Decays to two weak vector bosons, ZZ or WW, each of which decays to two fermions. Because the Z and W bosons' masses are larger than half of the Higgs boson's mass, the three particles involved in the $H \rightarrow VV$ decay cannot all be onshell. These decay can therefore be studied in two cases: the onshell region, where the Higgs boson has a mass of 125 GeV and one of the vector bosons has a mass lower than its pole mass, and the offshell region, where both vector bosons are onshell and the Higgs boson is offshell with a larger mass. The leptonic decays, particularly $H \rightarrow ZZ \rightarrow 4\ell$ and to a lesser extent $H \rightarrow WW \rightarrow 2\ell 2\nu$, are clean channels with relatively low backgrounds. In addition, the four-fermion system contains information about the polarization of the Z and W bosons, which can be used to study the tensor structure of the Higgs boson's couplings. This information is only available in decays such as this one, where four final state particles are involved.

The analyses presented here focus on the $H \rightarrow ZZ \rightarrow 4\ell$ and $H \rightarrow \tau\tau$ decays.

4.3 Mass and Width

The Standard Model does not predict any particular mass for the Higgs boson, so the mass has to be determined experimentally. Currently, the Particle Data Group's estimate for the Higgs boson's mass, based on a combination of CMS and ATLAS results, is 125.10 ± 0.14 GeV [7–10]. Once the mass is known, the Standard Model provides a full prediction for the rest of the Higgs boson's properties. For this reason, an experimental measurement of these properties constitutes a test of the Standard Model or, equivalently, a search for beyond Standard Model physics.

The simplest approach for measuring the Higgs boson's width is to simply measure the mass distribution of Higgs boson events and observe a Breit-Wigner distribution. However, for a mass of 125 GeV, the Standard Model predicts a width of (4.088 ± 0.056) MeV [1], orders of magnitude smaller than the resolution of any detectors targeting the Higgs boson. Therefore, although the simple procedure can test whether the Higgs boson width is larger than a GeV or so, there is not much hope that it will be able to approach the Standard Model expectation.

An alternative procedure is to look at offshell Higgs boson production. As mentioned earlier, at 125 GeV the Higgs boson cannot decay to two onshell Z bosons because the Z boson mass is too high. Above 200 GeV, an offshell Higgs boson can decay to two onshell Z bosons, and the Standard Model's prediction of the rate for this process is high enough to be observed at the LHC. The ratio between the onshell and offshell production rates is proportional to Higgs boson's width [11], and this can be used for a much more precise measurement of the Higgs boson's width.

4.4 Anomalous Couplings

The Standard Model predicts how the Higgs boson interacts with other particles. Starting shortly after the Higgs boson's discovery and continuing to the present, analyses by CMS [9, 12–20] and ATLAS [21–27] confirmed that the Higgs boson is a spin-0 particle and that it primarily interacts with vector bosons through the tree-level, scalar tensor structure predicted by the Standard Model. However, additional contributions to the Higgs boson's interaction are still possible. These anomalous couplings are small but nonzero even in the Standard Model through loop couplings, and beyond Standard Model effects can enhance anomalous couplings. Measurements of these anomalous couplings will be discussed in detail in Chap. 5.

4.4.1 Couplings to Vector Bosons

The possible interactions between the Higgs boson and other particles are limited by Lorentz invariance. In the case of its interaction with vector bosons VV, the only possible couplings up to $\mathcal{O}(q^2)$ are given by [28]:

$$A \sim \left[a_1^{\mathrm{VV}} - \frac{\kappa_1^{\mathrm{VV}} q_1^2 + \kappa_2^{\mathrm{VV}} q_2^2}{\left(\Lambda_1^{\mathrm{VV}}\right)^2} - \frac{\kappa_3^{\mathrm{VV}} (q_1 + q_2)^2}{\left(\Lambda_Q^{\mathrm{VV}}\right)^2} \right] m_{\mathrm{V1}}^2 \epsilon_{\mathrm{V1}}^* \epsilon_{\mathrm{V2}}^*$$
$$+ a_2^{\mathrm{VV}} f_{\mu\nu}^{*(1)} f^{*(2)\,\mu\nu} + a_3^{\mathrm{VV}} f_{\mu\nu}^{*(1)} \tilde{f}^{*(2)\,\mu\nu}, \tag{4.1}$$

where $f^{(i)\,\mu\nu} = \epsilon_i^\mu q_i^\nu - \epsilon_i^\nu q_i^\mu$ is the vector boson's field strength tensor, q_i and ϵ_i are its momentum and polarization, and $\tilde{f}^{(i)\,\mu\nu} = \frac{1}{2}\epsilon^{\mu\nu\alpha\beta} f_{\alpha\beta}^{(i)}$ is its conjugate field strength tensor. The couplings a_i^{VV} and $\frac{\kappa_i^{\mathrm{VV}}}{\left(\Lambda_{1,Q}^{\mathrm{VV}}\right)^2}$ are, in general, arbitrary complex numbers to be measured.

In the Standard Model, the only tree level coupling is $a_1^{\mathrm{ZZ}} = a_1^{\mathrm{WW}} = 2$. The Λ_1 and Λ_Q terms represent modifications to this term as a function of the momenta of the particles involved, either the vector bosons in the case of Λ_1 or the Higgs boson in the case of Λ_Q. Two other tensor structures, a_2 and a_3 are also possible. (They can also have q^2 modifiers in front of them, but these only enter at higher orders in q^2.) The a_1 and a_2 terms are CP-even, while a_3 is CP-odd.

This parameterization does not assume any particular mechanism for producing anomalous behavior. However, any model that predicts anomalous HVV couplings can be described through this parameterization (possibly with additional terms at higher order in q^2). For instance, a new Z′ boson with a mass much heavier than the Higgs boson can produce a finite Λ_1 term. A new, heavy t′ quark in a loop can

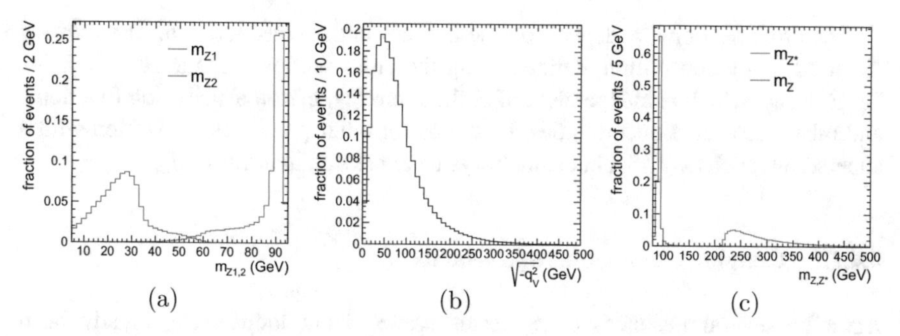

Fig. 4.4 Distributions of the vector boson mass for (**a**) the onshell Higgs boson decaying to four leptons, where m_1 is the dilepton mass closer to 91.2 GeV and m_2 is the other; (**b**) VBF, where q^2 is negative and therefore $\sqrt{-q^2}$ is plotted; and (**c**) ZH production, where Z^* decays to ZH

produce an a_2 term. This term is nonzero even in the Standard Model due to a t quark loop, but its expected value is too small to measure with current data.

Because the anomalous couplings terms are at higher order in q_V^2 than the Standard Model term, they become more important when more energy is exchanged. In the onshell H $\rightarrow$ ZZ decay, the two Z bosons typically have invariant masses of around 91.2 GeV and 40 GeV. The typical values of q^2 are larger in VBF and VH, as shown in Fig. 4.4. For this reason, small anomalous coupling will produce a greater effect in VBF and VH than in decay, and analyses targeting VBF and VH can achieve greater sensitivity to small anomalous couplings.

For the same reason, the anomalous couplings are enhanced in the offshell region, where the increased total invariant mass of the four-lepton system means that both Z bosons have an invariant mass of 91.2 GeV. The consequences are the same: an analysis targeting offshell events will be sensitive to smaller anomalous couplings than an analysis that only looks at the onshell region. The offshell region is also sensitive to the Λ_Q term, which depends *only* on the Higgs boson's invariant mass and is indistinguishable from a_1 in the onshell region.

It is convenient to measure the effective cross-section ratios (e.g. f_{ai}) rather than the anomalous couplings themselves (equivalently in the amplitude or EFT notation) because many systematic uncertainties cancel in the ratios and the physical range is bounded between 0 and 1. Moreover, these ratios are invariant with respect to the coupling convention. Therefore, our primary measurements are performed in the basis of cross-section ratios. For the electroweak vector boson couplings, the effective fractional ZZ cross sections f_{ai} and phases ϕ_{ai} are defined as

$$f_{ai} = \frac{|a_i|^2\sigma_i}{\sum_{j=1,2,3...}|a_j|^2\sigma_j}, \qquad \phi_{ai} = \arg\left(\frac{a_i}{a_1}\right), \tag{4.2}$$

where σ_i is the cross section for the process corresponding to $a_i = 1, a_{j\neq i} = 0$, while $\tilde{\sigma}_{\Lambda 1}$ is the effective cross section for the process corresponding to $\Lambda_1 = 1$ TeV, given in units of fb TeV ^{4}. The sum of all f_{ai}'s is always 1.

The cross sections σ_i depend on the process under consideration. The primary f_{ai} used is, by convention, defined using the cross sections for the $\mathrm{H} \to \mathrm{ZZ} \to 2\mathrm{e}2\mu$ decay, which is independent of collider energy, parton distribution functions, and other associated uncertainties. Fractions for other processes are written with a superscript, such as f_{ai}^{VBF}, but can always be expressed in terms of f_{ai}.

4.4.1.1 Couplings to Photons and Gluons

Actually, several instances of a_2 terms arising from loops have already been observed. In the case of $\mathrm{H} \to \gamma\gamma$ couplings, gauge invariance forbids the first tensor structure; the only possibilities remaining are $a_2^{\gamma\gamma}$ and $a_3^{\gamma\gamma}$. One of these tensor structures must be involved in the diphoton decay as observed by CMS and ATLAS. Based on the Standard Model, the t and W loops should produce an a_2 tensor structure, and the a_3 term should be highly suppressed. This cannot be directly confirmed experimentally because the helicity information carried by the photons is lost. With more data, it will become possible to distinguish between $a_2^{\gamma\gamma}$ and $a_3^{\gamma\gamma}$ in $\mathrm{H} \to \gamma^*\gamma^* \to 4f$ decays, where this information is available; however, the measured rate of $\mathrm{H} \to \gamma\gamma$ indicates that this process is currently too rare to detect. (Other processes that involve Htt and HWW couplings can also shed light on the nature of $\mathrm{H} \to \gamma\gamma$ couplings, albeit indirectly.)

The same applies to the Hgg coupling, which is also expected in the Standard Model to proceed through a t quark loop that gives rise to an a_2^{gg} term. The bare gluon fusion process shown in Fig. 4.1a does not provide enough information to distinguish between a_2^{gg} and a_3^{gg}. However, if there are two or more extra jets in the final state, their angular correlations carry information that can be used to measure the tensor structure of the Hgg interactions. One diagram, which is similar to VBF (Fig. 4.1b), is illustrated in Fig. 4.5. The amount of data collected is just sufficient for first measurements of this process, which will be discussed later. For the gluon fusion process, we use, from Eq. (4.2):

$$f_{a3}^{\mathrm{ggH}} = \frac{|a_3^{\mathrm{gg}}|^2}{|a_2^{\mathrm{gg}}|^2 + |a_3^{\mathrm{gg}}|^2} \tag{4.3}$$

where σ_2^{gg} and σ_3^{gg} are equal and therefore drop out in the ratio.

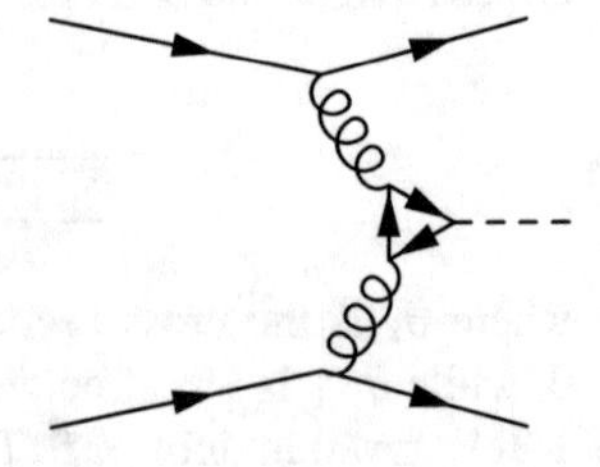

Fig. 4.5 One of the Feynman diagrams for gluon fusion production with two extra jets in the final state

The $HZ\gamma$ couplings are also expected to proceed through a t or W loop that forms an effective $a_2^{Z\gamma}$ coupling. So far, the $\mathrm{H} \to \mathrm{Z}\gamma \to 2f\gamma$ process has not been observed, and limits on its production cross section indicate that $\mathrm{Z}\gamma^*$ contributions to the $\mathrm{H} \to 4f$ final state are still out of reach, assuming that this process proceeds through the $a_2^{Z\gamma}$ or $a_3^{Z\gamma}$ couplings.

The $\Lambda_1^{Z\gamma}$ coupling is not, and cannot be, excluded by direct measurements of $\mathrm{H} \to \mathrm{Z}\gamma \to 2f\gamma$. Gauge invariance requires that $\kappa_1^{Z\gamma} = 0$ in Eq. (4.1), so that the contribution for this term is $A_{\Lambda 1}^{Z\gamma} \sim \frac{q_\gamma^2}{\left(\Lambda_1^{Z\gamma}\right)^2} m_Z^2 \epsilon_Z^* \epsilon_\gamma^*$. The q_γ^2 term in the numerator sets this contribution to 0 for any process involving an onshell photon. Therefore, the only way to measure this coupling is through processes where its contribution would involve an offshell photon, such as $\mathrm{H} \to \mathrm{Z}\gamma^* \to 4f$, and these measurements form a part of the $\mathrm{H} \to \mathrm{ZZ} \to 4\ell$ analysis.

4.4.2 Couplings to Fermions

In the case of the Higgs boson's couplings to a fermion field ψ_f with mass m_f, there are only two possible tree-level terms in the amplitude [29]:

$$A = -\frac{m_\mathrm{f}}{v} \bar{\psi}_\mathrm{f} \left(\kappa_\mathrm{f} + i\tilde{\kappa}_\mathrm{f}\gamma_5\right) \psi_\mathrm{f} \tag{4.4}$$

The first term gives a scalar coupling, while the second gives a pseudoscalar coupling. In the Standard Model, $\kappa = 1$ and $\tilde{\kappa} = 0$.

Similarly to f_{ai} for HVV couplings, we define

$$f_{CP}^{\mathrm{Hff}} = \frac{|\tilde{\kappa}_\mathrm{f}|^2}{|\kappa_\mathrm{f}|^2 + |\tilde{\kappa}_\mathrm{f}|^2} \tag{4.5}$$

to parameterize the fraction of pseudoscalar Hff couplings. Although the cross sections are in general different, we drop the cross section ratio in this definition to remove dependence on uncertainties related to production cross sections.

These couplings are more difficult to measure than the HVV couplings. Many of the common processes involving Hff couplings do not carry information sensitive to the precise structure of those couplings. For example, the only way to extract this information from the $\mathrm{H} \to \mathrm{b\bar{b}}$ and $\mathrm{H} \to \tau\tau$ decays is through the helicity of the τ leptons, which is possible [29] but made more difficult by detector resolution, or b quarks, which is even more difficult due to QCD effects. However, the rare $\mathrm{t\bar{t}H}$ process is sensitive to Htt couplings. Also, under the assumption that gluon fusion proceeds through a top and bottom loop with $\kappa_t = \kappa_b$ and $\tilde{\kappa}_t = \tilde{\kappa}_b$

$$\left|f_{CP}^{\mathrm{Hff}}\right| = \left(1 + 2.38\left[\frac{1}{\left|f_{a3}^{\mathrm{ggH}}\right|} - 1\right]\right)^{-1} = \sin^2\alpha^{\mathrm{Hff}}, \tag{4.6}$$

where the signs of f_{CP}^{Hff} and f_{a3}^{ggH} are equal, and α^{Hff} is an effective parameter sometimes used to describe the CP-odd contribution to the Higgs boson's Yukawa couplings. These couplings can be measured by studying gluon fusion production with two associated jets, as described in the previous section.

The first measurements of Htt couplings in these two processes will be discussed in Sect. 5.7.

4.4.3 Effective Field Theory

The parameterization of the amplitude in Eq. (4.1) can be related to a fundamental Lagrangian density function, using effective field theory (or EFT) coefficients of the so-called Higgs basis [1]:

$$\begin{aligned}
\mathcal{L}_{\mathrm{hvv}} = \frac{h}{v}\Bigg[& (1+\delta c_z)\frac{(g^2+g'^2)v^2}{4}Z_\mu Z_\mu + c_{zz}\frac{g^2+g'^2}{4}Z_{\mu\nu}Z_{\mu\nu} \\
& + c_{z\Box}g^2 Z_\mu \partial_\nu Z_{\mu\nu} + \tilde{c}_{zz}\frac{g^2+g'^2}{4}Z_{\mu\nu}\tilde{Z}_{\mu\nu} \\
& + (1+\delta c_w)\frac{g^2v^2}{2}W_\mu^+ W_\mu^- + c_{ww}\frac{g^2}{2}W_{\mu\nu}^+ W_{\mu\nu}^- \\
& + c_{w\Box}g^2\left(W_\mu^- \partial_\nu W_{\mu\nu}^+ + \mathrm{h.c.}\right) + \tilde{c}_{ww}\frac{g^2}{2}W_{\mu\nu}^+\tilde{W}_{\mu\nu}^- \\
& + c_{z\gamma}\frac{e\sqrt{g^2+g'^2}}{2}Z_{\mu\nu}A_{\mu\nu} + \tilde{c}_{z\gamma}\frac{e\sqrt{g^2+g'^2}}{2}Z_{\mu\nu}\tilde{A}_{\mu\nu} + c_{\gamma\Box}gg'Z_\mu\partial_\nu A_{\mu\nu} \\
& + c_{\gamma\gamma}\frac{e^2}{4}A_{\mu\nu}A_{\mu\nu} + \tilde{c}_{\gamma\gamma}\frac{e^2}{4}A_{\mu\nu}\tilde{A}_{\mu\nu} + c_{gg}\frac{g_s^2}{4}G_{\mu\nu}^a G_{\mu\nu}^a + \tilde{c}_{gg}\frac{g_s^2}{4}G_{\mu\nu}^a\tilde{G}_{\mu\nu}^a\Bigg],
\end{aligned} \tag{4.7}$$

in accordance with Eq. (II.2.20) in Ref. [1].

There is a unique representation of each EFT coefficient in Eq. (4.7) by couplings in Eq. (4.1), as shown below [30]:

$$\delta c_z = \frac{1}{2}a_1^{ZZ} - 1, \quad c_{zz} = -\frac{2s_w^2c_w^2}{e^2}a_2^{ZZ}, c_{z\Box} = \frac{M_Z^2 s_w^2}{e^2}\frac{\kappa_1^{ZZ}}{(\Lambda_1^{ZZ})^2}, \quad \tilde{c}_{zz} = -\frac{2s_w^2c_w^2}{e^2}a_3^{ZZ},$$

$$\delta c_w = \frac{1}{2}a_1^{WW} - 1, \quad c_{ww} = -\frac{2s_w^2}{e^2}a_2^{WW}, c_{w\Box} = \frac{M_W^2 s_w^2}{e^2}\frac{\kappa_1^{WW}}{(\Lambda_1^{WW})^2}, \quad \tilde{c}_{ww} = -\frac{2s_w^2}{e^2}a_3^{WW},$$

$$c_{z\gamma} = -\frac{2s_w c_w}{e^2} a_2^{Z\gamma}, c_{\gamma\Box} = \frac{s_w c_w}{e^2} \frac{M_Z^2}{(\Lambda_1^{Z\gamma})^2} \kappa_2^{Z\gamma}, \quad \tilde{c}_{z\gamma} = -\frac{2s_w c_w}{e^2} a_3^{Z\gamma},$$
$$c_{\gamma\gamma} = -\frac{2}{e^2} a_2^{\gamma\gamma}, \qquad \tilde{c}_{\gamma\gamma} = -\frac{2}{e^2} a_3^{\gamma\gamma},$$
$$c_{gg} = -\frac{2}{g_s^2} a_2^{gg}, \qquad \tilde{c}_{gg} = -\frac{2}{g_s^2} a_3^{gg}. \tag{4.8}$$

Enforcing linear relations for the dependent coefficients [1] allows a unique relationship based on a minimal set of degrees of freedom [30]:

$$a_1^{WW} = a_1^{ZZ}, \tag{4.9}$$

$$a_2^{WW} = c_w^2 a_2^{ZZ} + s_w^2 a_2^{\gamma\gamma} + 2s_w c_w a_2^{Z\gamma}, \tag{4.10}$$

$$a_3^{WW} = c_w^2 a_3^{ZZ} + s_w^2 a_3^{\gamma\gamma} + 2s_w c_w a_3^{Z\gamma}, \tag{4.11}$$

$$\frac{\kappa_1^{WW}}{(\Lambda_1^{WW})^2}(c_w^2 - s_w^2) = \frac{\kappa_1^{ZZ}}{(\Lambda_1^{ZZ})^2}, +2s_w^2 \frac{a_2^{\gamma\gamma} - a_2^{ZZ}}{M_Z^2} + 2\frac{s_w}{c_w}(c_w^2 - s_w^2)\frac{a_2^{Z\gamma}}{M_Z^2}, \tag{4.12}$$

$$\frac{\kappa_2^{Z\gamma}}{(\Lambda_1^{Z\gamma})^2}(c_w^2 - s_w^2) = 2s_w c_w \left(\frac{\kappa_1^{ZZ}}{(\Lambda_1^{ZZ})^2} + \frac{a_2^{\gamma\gamma} - a_2^{ZZ}}{M_Z^2}\right) + 2(c_w^2 - s_w^2)\frac{a_2^{Z\gamma}}{M_Z^2}. \tag{4.13}$$

4.5 Simulation

Several programs are used to simulate Higgs boson events at the LHC for the analyses described here.

4.5.1 JHUGEN

The JHU generator [28–32], or JHUGEN, generates a spin 0, 1, or 2 particle under a generic coupling model. Since the Higgs boson's discovery, development has naturally focused on understanding this particle and exploring more of its possible interactions, while also keeping the flexibility to search for and characterize other new resonances.

For a spin 0 particle, JHUGEN supports all of the processes described in Sects. 4.1 and 4.2. Within those processes, the HVV coupling model includes Eq. (4.1), where a_1^{VV}, a_2^{VV}, a_3^{VV}, and Λ_1^{VV} can be set to arbitrary complex numbers, as can several higher order terms. Scalar and pseudoscalar couplings to fermions, κ and $\tilde{\kappa}$, can also be set for processes involving Hff couplings.

JHUGEN produces either weighted events in the form of histograms or, more relevant for the analyses here, unweighted events in the Les Houches, or LHE, format [33, 34].

JHUGEN is used extensively for all of the analysis work shown here. Figures 4.4, 4.9 and 4.10 show some distributions produced by JHUGEN.

4.5.2 *MCFM and MCFM*+JHUGEN

The MCFM generator [35] is used to generate the offshell tail of the Higgs boson, including its interference with background. MCFM is used directly to generate gluon fusion, gg → ZZ background, and their interference. Modifications to the MCFM source code, provided by JHUGEN, enable all of the anomalous couplings to be used for offshell generation. An additional spin-0 resonance, which interferes with the H boson and with background, can also be included. Mass distributions are shown in Fig. 4.6.

In addition, JHUGEN wraps the MCFM matrix element to generate offshell VBF and VH, vector boson scattering (VBS) and tri-vector-boson production (VVV), and their interference, including anomalous couplings. The JHUGEN code enables generation of LHE events for this process, which MCFM does not include. This is the most complicated process implemented in JHUGEN to date, and it uses several optimizations to make generation more efficient.

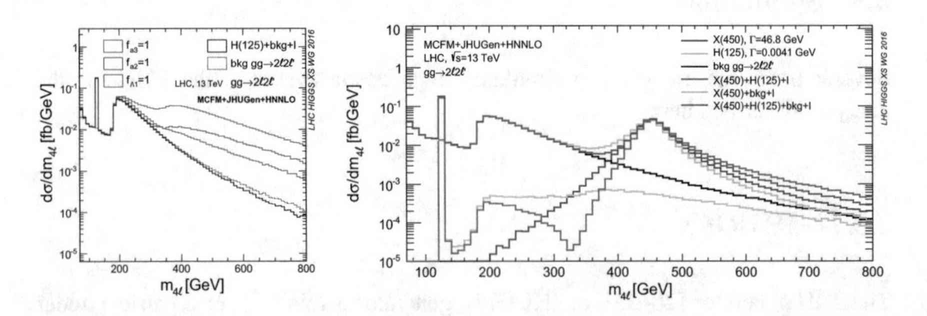

Fig. 4.6 Distribution of $m_{4\ell}$ for gg → 4ℓ, including background (bkg) and the Higgs boson (H). The left plot includes several anomalous coupling hypotheses for the H boson. The right plot also includes an additional spin-0 resonance with $m_X = 450$ GeV, $\Gamma_X = 46.8$ GeV [1]

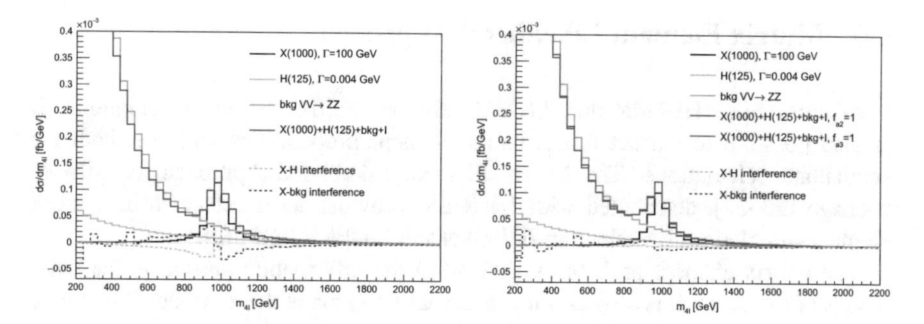

Fig. 4.7 Differential cross section of the process VV $\rightarrow$ ZZ/Z$\gamma^*/\gamma^*\gamma^*$ $\rightarrow$ $2\ell 2\ell'$ (where ℓ, ℓ' = e, μ, or τ) as a function of invariant mass $m_{4\ell}$ generated with JHUGEN, including VBS and VVV background (bkg) and VBF+VH production of both the Higgs boson (H) and a hypothetical resonance X(1000). The contributions to the final differential cross section are shown isolated and combined. In all cases interference (I) of all contributing amplitudes is included. The left panel shows the scalar X hypothesis, and the right panel shows the XVV $f_{a2} = 1$ and $f_{a3} = 1$ hypotheses [30]

Figure 4.7 shows distributions of $m_{4\ell}$ for a new resonance with a mass of 1 TeV with several signal, background, and interference components separated. Any search for this kind of resonance must take all of these components into account in determining the final distribution of mass and kinematics. Additionally, a new resonance will not necessarily interact with other particles with the same tensor structure as the Higgs boson. Its coupling structure can significantly change the final mass shape, as the figures show.

4.5.3 Higher Order Corrections

JHUGEN and MCFM are leading order generators. In the gluon fusion process, production at next-to-leading order (NLO), with a single extra jet, is unaffected by anomalous couplings. POWHEG [36, 37], which is an NLO generator for Standard Model processes, is therefore used to generate the gluon fusion production mechanism to include NLO effects. JHUGEN is still used for the H $\rightarrow$ VV $\rightarrow$ 4ℓ decay.

For other processes, POWHEG is used to generate the Standard Model hypothesis at next-to-leading order [38, 39]. For gluon fusion with two extra jets and for VH, the MINLO-HJJ [40] and MINLO-HVJ [41] extensions are used. The results are compared to JHUGEN's Standard Model simulation, and the differences in relevant kinematics are small. This sets the scale for potential differences between JHUGEN's simulation of anomalous hypotheses and any NLO corrections that may arise there.

4.6 Matrix Element Likelihood Approach

Generators like JHUGEN and MCFM rely on matrix element calculations. It is also possible to extract those matrix element probabilities and use them for standalone calculations. The Matrix Element Likelihood Approach, or MELA, package [28–32], distributed with JHUGEN, provides an interface to the matrix element calculations contained in JHUGEN and MCFM+JHUGEN.

The matrix element probability uses all of the information present in the kinematics of the event. It is also useful conceptually to look at the individual kinematic distributions that go into this calculation. A general system of four fermion momenta has 16 components. Of these, four are fixed by the fermion masses, which can be approximated as zero. Two more can be absorbed by a physically uninteresting rotation around the z axis and boost along the z axis, keeping the beamline fixed. Because MELA uses a leading order calculation, an additional two components are eliminated by setting the system's transverse momentum $\vec{p}_T = (p_x, p_y)$ to zero.

The system of particles involved in the four-fermion decay or in VBF or VH production can therefore be fully described by eight independent parameters. The parameterization used here includes (1) the Higgs boson's mass $m_{4\ell}$; (2-3) the two vector boson masses $m_{1,2}$; (4-6) four angles describing the fermion kinematics, $\theta_{1,2}$ and Φ; and (7-8) two angles relating production to decay, θ^* and Φ_1. These observables are illustrated in Fig. 4.8, together with a similar but more complicated set of angles that describe the $\mathrm{t\bar{t}H}$ process.

For an onshell Higgs boson, $m_{4\ell} = 125$ GeV to well within the detector precision. Also, for a spin-zero particle, the distributions of θ^* and Φ_1 are flat. These three parameters are useful to separate signal from background, and the angles were also used in [14] as part of the analysis to exclude higher spin hypotheses. However, they cannot distinguish between different spin 0 coupling hypotheses.

Distributions of the five remaining observables for the $\mathrm{H} \to \mathrm{ZZ} \to 4\ell$, VBF, and VH processes for a variety of coupling hypotheses are shown in Figs. 4.9 and 4.10.

The matrix elements calculated by MELA have two primary uses: to reweight generated events from one hypothesis to another and to create kinematic discriminants that can be used to separate hypotheses.

4.6.1 *Reweighting*

The driving principle behind Monte Carlo generators is that, in a sample generated to simulate a hypothesis a, the number of events in the output sample in a particular phase space region, $d\vec{\Omega}$ around $\vec{\Omega}$, is proportional to the probability to obtain an event in $d\vec{\Omega}$ using a particular probability distribution for that hypothesis P^a.

$$N^a(\vec{\Omega})d\vec{\Omega} \sim P^a(\vec{\Omega})d\vec{\Omega} \tag{4.14}$$

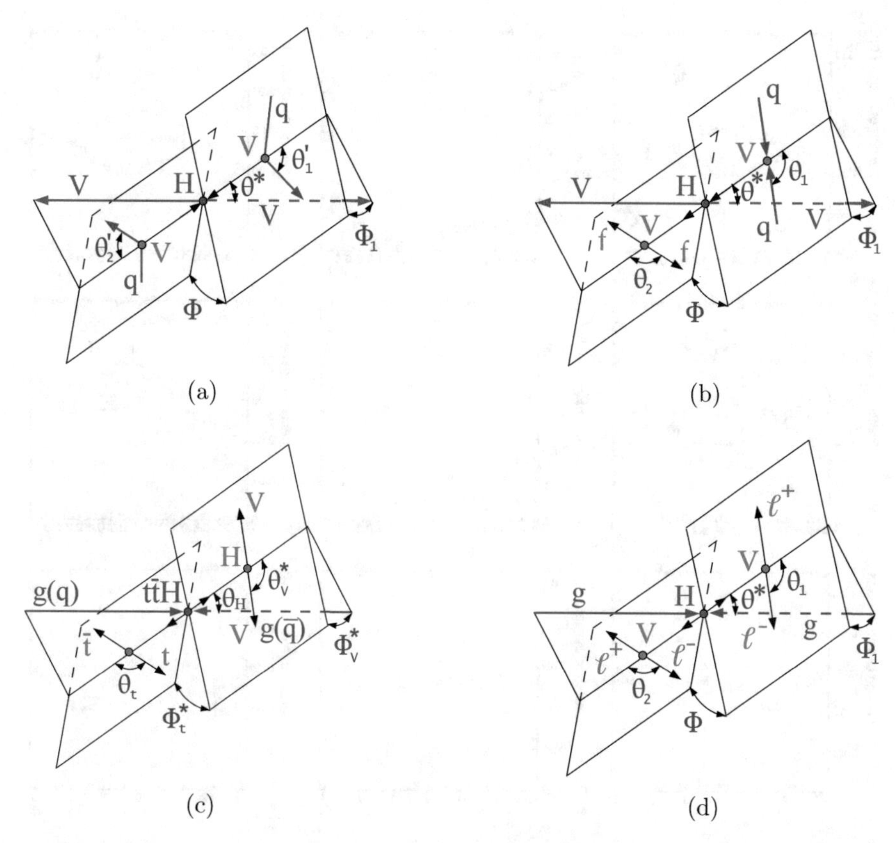

Fig. 4.8 Four topologies of the H boson production and decay: vector boson fusion qq → VV(qq) → H(qq) → VV(qq) (top-left); associated production qq → V → VH → (ff̄) H → (ff̄) VV (top-right); tt̄H or tqH production in association with the top quarks (bottom-left); and gluon fusion with decay gg → H → VV → 4ℓ (bottom-right), representing the topology without associated particles. The incoming particles are shown in brown, the intermediate vector bosons and their fermion daughters are shown in green, the H boson and its vector boson daughters are shown in red, and angles are shown in blue. In the first three cases the production and decay H → VV are followed by the same four-lepton decay shown in the last case. The angles are defined in either the respective rest frames [28, 29, 31], and subsequent top quark decay is not shown, but could also be included [29]

(If everything is done correctly, the probability distribution matches the actual probability distribution of events being studied, which in our case means collisions at the LHC. JHUGEN provides a good approximation. However, this discussion of reweighting is purely mathematical and does not rely on how accurate the simulation is.)

If we index the events by i and assign a weight $w_i^a = 1$ to each event, Eq. (4.14) becomes

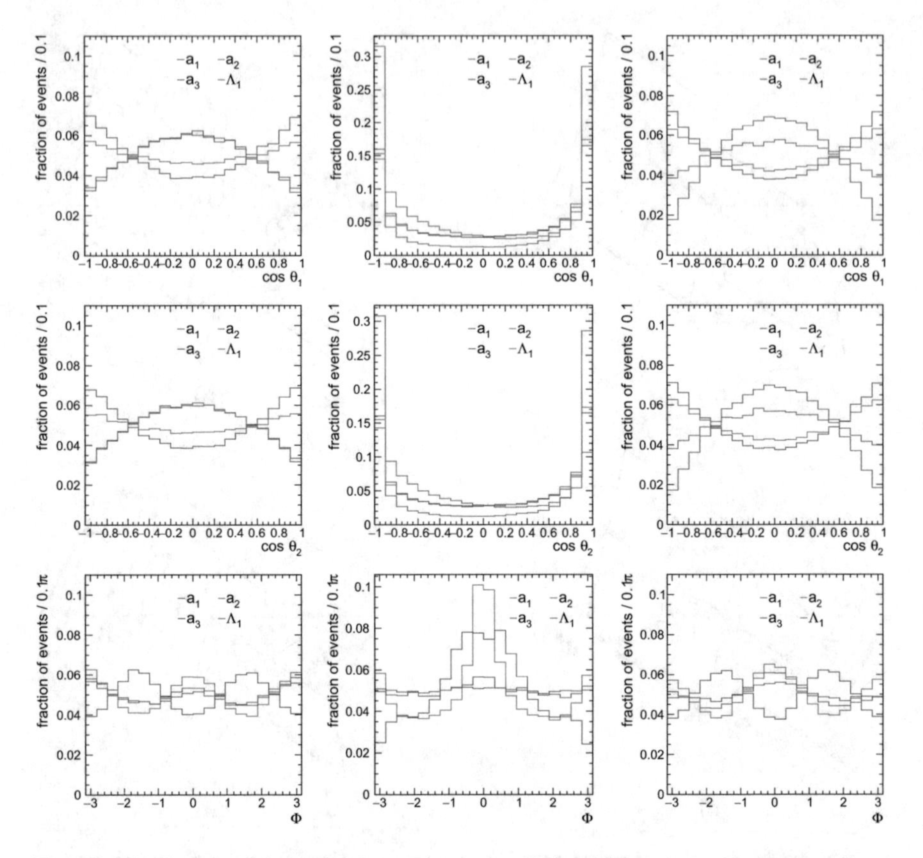

Fig. 4.9 Distributions of the angles that can be used to distinguish between anomalous couplings. The left column shows the distributions for the H → ZZ → 4ℓ decay, the middle column shows the distributions for VBF production, and the right column shows the distributions for VH production. From top to bottom, the distributions shown are $\cos\theta_1$, $\cos\theta_2$, ϕ. The Standard Model hypothesis a_1 and three anomalous hypotheses a_2, a_3 and Λ_1 are shown

$$\left(\sum_{\vec{\Omega}_i \in d\vec{\Omega}} w_i^a\right) d\vec{\Omega} \sim P^a(\vec{\Omega})d\vec{\Omega} \tag{4.15}$$

Next, we assign a second weight to each event in the sample, corresponding to a second hypotheses b:

$$w_i^b = \frac{P^b\left(\vec{\Omega}_i\right)}{P^a\left(\vec{\Omega}_i\right)} w_i^a \tag{4.16}$$

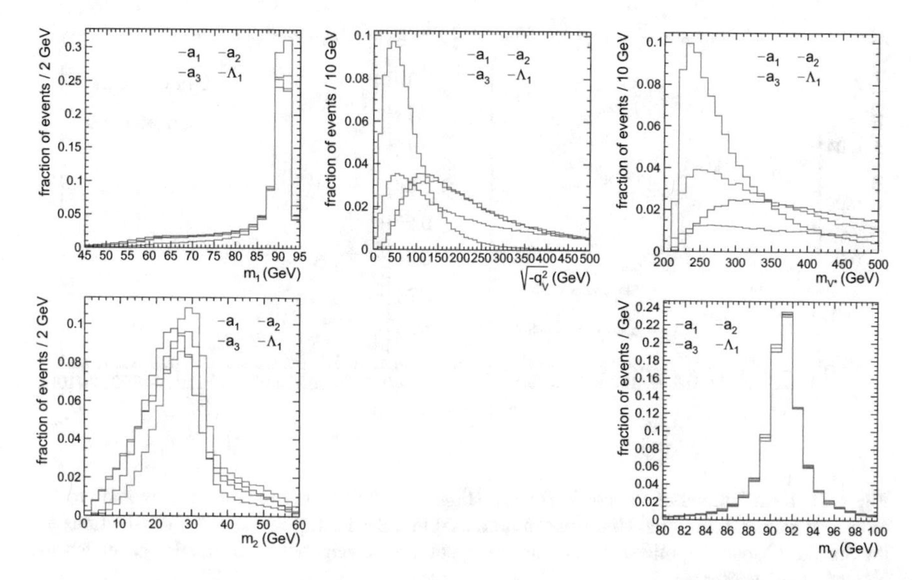

Fig. 4.10 Distributions of the invariant masses that can be used to distinguish between anomalous couplings. The left column shows the distributions for the H $\to$ ZZ $\to 4\ell$ decay, where m_1 is defined as the heavier dilepton system's mass and m_2 is the lighter one. The middle column shows the distribution for VBF production. The right column shows the distributions for VH production. The Standard Model hypothesis a_1 and three anomalous hypotheses a_2, a_3 and Λ_1 are shown

At any particular phase space point $\vec{\Omega}$,

$$\begin{aligned}\left(\sum_{\vec{\Omega}_i \in d\vec{\Omega}} w_i^b\right) d\vec{\Omega} &= \left(\sum_{\vec{\Omega}_i \in d\vec{\Omega}} w_i^a\right) \frac{P^b(\vec{\Omega})}{P^a(\vec{\Omega})} d\vec{\Omega} \\ &\sim P^a(\vec{\Omega}) \frac{P^b(\vec{\Omega})}{P^a(\vec{\Omega})} d\vec{\Omega} \\ &= P^b(\vec{\Omega}) d\vec{\Omega}\end{aligned} \tag{4.17}$$

The two ends of Eq. (4.17) look exactly like Eq. (4.14). We now have a way to use the same exact events to simulate two different hypotheses by using two sets of weights. This procedure relies on the capability, provided by MELA, to calculate P^a and P^b for a particular event, independent of the generator.

With an infinite sample, there would be no limitations on reweighting. In any real case, a limited number of events are generated, and reweighting reduces the *effective* number of events even further. The severity of this reduction depends on how similar hypotheses a and b are. If there is a region where b's probability density is high but a's is low, then that region will be populated by fewer events in the sample reweighted from a, and the statistical precision will be poor. This is illustrated in

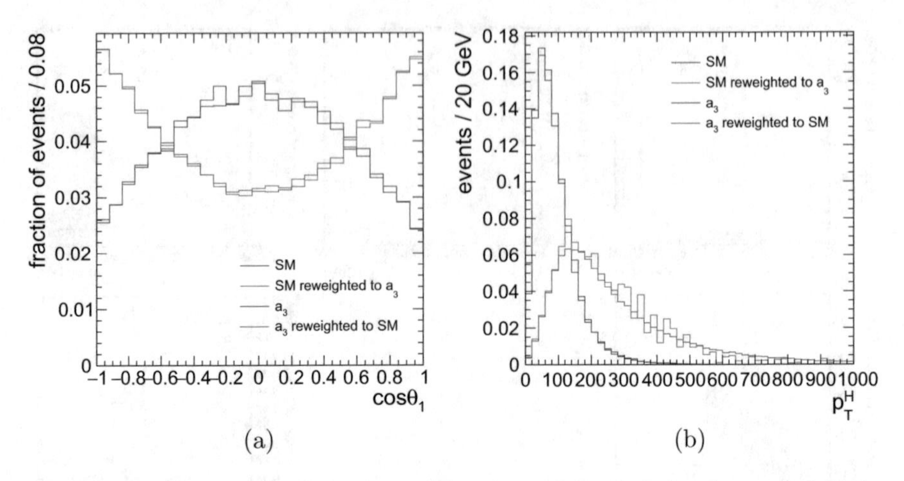

Fig. 4.11 Distributions of (**a**) $\cos\theta_1$ for the Higgs boson decaying to four leptons and (**b**) the transverse momentum of the Higgs boson produced in VBF. Each plot shows the distributions for the Standard Model hypothesis and a pure a_3 hypothesis reweighted from samples generated for both of these hypotheses

Fig. 4.11. In (a), the distributions for the two hypotheses are similar enough that the reweighting produces a reasonably smooth distribution, while in (b) the differences are large enough that the reweighted distributions, though correct on average, are not particularly useful in representing the hypotheses in question in the tail.

Therefore, it is necessary to generate a variety of samples, so that between them the whole phase space is covered. However, we then reweight each sample to each other sample, taking care to avoid the spikes. One procedure used to use all of the samples while minimizing the effect of noise will be discussed in detail in Sect. 5.6.2.

4.6.2 Discriminants

A second use for matrix element calculations is to create discriminants to distinguish between different hypotheses. A simple analysis can involve a collection of events, characterized by measured variables $\vec{\Omega}$, that might have been produced through one of two processes a or b. We want to determine which of those hypotheses is correct. By the Neyman-Pearson lemma [42], the best way to do this is through the ratio of probabilities to produce an event at $\vec{\Omega}$ under hypotheses a and b:

$$d_{ab}(\vec{\Omega}) = \frac{p_a(\vec{\Omega})}{p_b(\vec{\Omega})} \tag{4.18}$$

No other observable can help further: all the information carried by Ω that is useful to separate between a and b is contained in this discriminant.

Equivalently, we can use a function of this ratio, and we typically use

$$\mathcal{D}_{ab} = \frac{p_a}{p_a + p_b} = \frac{1}{1 + 1/d_{ab}}, \tag{4.19}$$

which has the advantage of being bounded between 0 and 1. This bounding is important because in practice, we have to bin the discriminant distribution, and in the process some information, corresponding to changes in shape within a single bin, is lost. The $\mathcal{D}_{ab}$ formulation mitigates this effect.

If there is a possibility of a non-interfering mixture between a and b, $\mathcal{D}_{ab}$ is still the best observable to distinguish between them. This is because any intermediate hypotheses c and d can be parameterized by $0 \le f_a^{c,d} \le 1$ and their probabilities can be expressed as

$$\begin{aligned} p_c &= f_a^c p_a + (1 - f_a^c) p_b \\ p_d &= f_a^d p_a + (1 - f_a^d) p_b \\ d_{cd} &= \frac{f_a^c p_a + (1 - f_a^c) p_b}{f_a^d p_a + (1 - f_a^d) p_b} \\ &= \frac{f_a^c d_{ab} + (1 - f_a^c)}{f_a^d d_{ab} + (1 - f_a^d)} \end{aligned} \tag{4.20}$$

which, again, is just a function of d_{ab}

When there are *three* separate processes a, b, and c, the two-dimensional distribution of two discriminants d_{ab} and d_{ac} is needed to obtain the full information to separate them. The third discriminant d_{bc} would be redundant because it is a function of the first two. Similarly, separating n processes requires $n - 1$ discriminants.

An important special case is two processes that interfere with each other. The probability for an intermediate hypothesis c, with constants g_a and g_b multiplying the amplitudes A_a and A_b, is

$$\begin{aligned} p_c &= |g_a A_a + g_b A_b|^2 \\ &= |g_a|^2 p_a + |g_b|^2 p_b + 2\mathrm{Re}(g_a^* g_b A_a^* A_b) \\ &= |g_a|^2 p_a + |g_b|^2 p_b + 2\mathrm{Re}(g_a^* g_b)\mathrm{Re}(A_a^* A_b) - 2\mathrm{Im}(g_a^* g_b)\mathrm{Im}(A_a^* A_b) \\ &= (\cdots) p_a + (\cdots) p_b + (\cdots) p_{\text{int}} + (\cdots) p_{int}^{\perp} \end{aligned} \tag{4.21}$$

For the purpose of discriminants, this mixture can be treated as a sum of four processes. The analyses described here assume that the couplings g_a and g_b are always real, so that there are only three terms, p_a, p_b, and p_{int}, and only two discriminants are needed. The discriminants we usually choose are $\mathcal{D}_{ab}$ and

$$\mathcal{D}_{\text{int}}^{ab} = \frac{p_{\text{int}}}{2\sqrt{p_a p_b}} \tag{4.22}$$

$$\text{or: } \mathcal{D}_{\text{int}}^{\prime ab} = \frac{p_{\text{int}}}{p_a + p_b} \tag{4.23}$$

This formulation of $\mathcal{D}_{\text{int}}$ is typically almost orthogonal to $\mathcal{D}_{ab}$ and is bounded between -1 and 1 for any value of $\mathcal{D}_{ab}$. ($\mathcal{D}_{\text{int}}^{\prime}$ was used instead in some older papers.)

4.6.2.1 Discriminant Examples: VBF and ggH

In this section [30], we illustrate the power of the matrix element technique in application to both VBF and ggH with two jets. In VBF, for illustration purposes we consider equal strength of WW and ZZ fusion with $a_1^{\text{ZZ}} = a_1^{\text{WW}}$ and $a_3^{\text{ZZ}} = a_3^{\text{WW}}$ in Eq. (4.1) and vary the relative contribution of the CP-even and CP-odd amplitudes, with the f_{a3}^{VBF} parameter representing their relative cross section fraction. The relative strength of the WW and ZZ fusion is fixed in this study because the two processes are essentially indistinguishable in their observed kinematics. In the strong boson fusion, the parameter f_{a3}^{ggH} represents a similar relative cross section fraction of the pseudoscalar coupling component.

Figure 4.12 shows distributions of the $\mathcal{D}_{0-}$ and $\mathcal{D}_{CP}$ discriminants, calculated according to Eqs. (4.19) and (4.22) for the VBF process, to distinguish between the SM hypothesis $a_1^{\text{ZZ}} = a_1^{\text{WW}} = 2$, the alternative hypothesis $a_3^{\text{ZZ}} = a_3^{\text{WW}} \neq 0$, and the interference between these two contributions. Figure 4.13 shows the same type of discriminants defined and shown for the ggH process, enhanced with the events in the VBF-like topology using the requirement $m_{JJ} > 300$ GeV for illustration. The

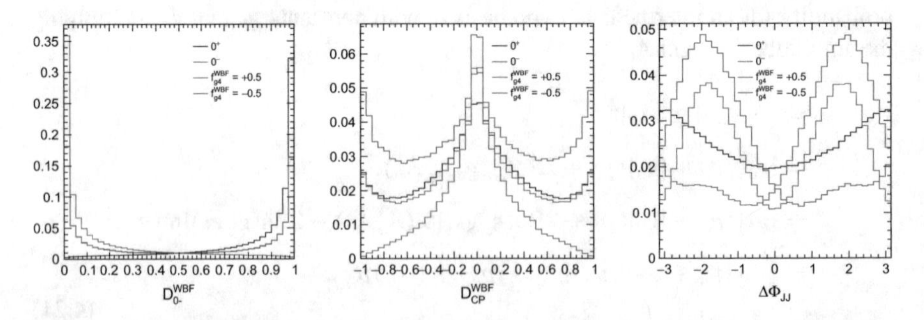

Fig. 4.12 Two discriminants defined in Eq. (4.19) (left) and Eq. (4.22) (middle) for the measurement of the CP-sensitive parameter f_{a3}^{VBF} in VBF production. Also shown is the $\Delta\Phi_{JJ}$ observable (right). The values of $f_{a3}^{\text{VBF}} = \pm 0.5$ correspond to 50% mixture of the CP-even and CP-odd contributions, where the couplings have opposite signs in the case of the negative value [30]

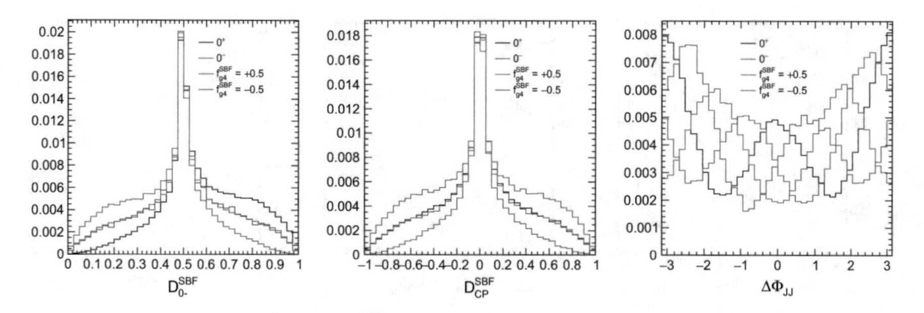

Fig. 4.13 Two discriminants defined in Eq. (4.19) (left) and Eq. (4.22) (middle) for the measurement of the CP-sensitive parameter f_{a3}^{ggH} in ggH production. Also shown is the $\Delta\Phi_{JJ}$ observable (right). The values of $f_{a3}^{\mathrm{ggH}} = \pm 0.5$ correspond to 50% mixture of the CP-even and CP-odd contributions, where the couplings have opposite signs in the case of the negative value. A requirement $m_{JJ} > 300$ GeV is applied to enhance the VBF-like topology of events [30]

latter requirement is based on the following observation. Among the initial states in the ggH process, we could have gg, qg, and qq parton pairs. The events with the quark-quark initial state carry most of the information for CP measurements and have the topology most similar to VBF process, which is also known to have large di-jet invariant mass. In both the VBF and ggH cases, the azimuthal angle difference between the two jets $\Delta\Phi_{JJ}$ is also shown for comparison [43]. It is similar to the Φ^{VBF} angle defined in Fig. 4.8 and shown in Fig. 4.9, but differs somewhat due to different frames used in the angle definition.

The $\Delta\Phi_{JJ}$ angle is defined as follows. The directions of the two jets are represented by the vectors $\vec{j}_{1,2}$ in the laboratory frame, and $\vec{j}_{T1,2}$ are the transverse components in the xy plane. If we label j_1 as the jet going in the $-z$ direction (or less forward) and j_2 as the jet going in the $+z$ direction (or more forward), then $\Delta\Phi_{JJ}$ is the azimuthal angle difference between the first and the second jets, or $\phi_1 - \phi_2$. In vector notation,

$$\Delta\Phi_{JJ} = \frac{(\hat{j}_{T1} \times \hat{j}_{T2}) \cdot \hat{z}}{|(\hat{j}_{T1} \times \hat{j}_{T2}) \cdot \hat{z}|} \cdot \frac{(\vec{j}_1 - \vec{j}_2) \cdot \hat{z}}{|(\vec{j}_1 - \vec{j}_2) \cdot \hat{z}|} \cdot \cos^{-1}\left(\hat{j}_{T1} \cdot \hat{j}_{T2}\right), \tag{4.24}$$

where the angle between $\vec{j}_{T1}$ and $\vec{j}_{T2}$ defines $\Delta\Phi_{JJ}$ and the two ratios provide the sign convention. This definition is invariant under the exchange of the two jets or the choice of the positive z axis direction.

The information content of the observables can be illustrated with the Receiver Operating Characteristic (ROC) curve, which is a graphical plot that illustrates the diagnostic ability of a binary classifier system as its discrimination threshold is varied. Figure 4.14 (left) shows the ROC curves illustrating discrimination between scalar and pseudoscalar models in the VBF process using the $\mathcal{D}_{0-}$ and $\Delta\Phi_{JJ}$ observables. The optimal observable $\mathcal{D}_{0-}$, which incorporates all kinematic and dynamic information, visible in multiple individual observables shown in Figs. 4.10 and 4.9, has the clear advantage. Figure 4.14 (right) shows the same comparison in

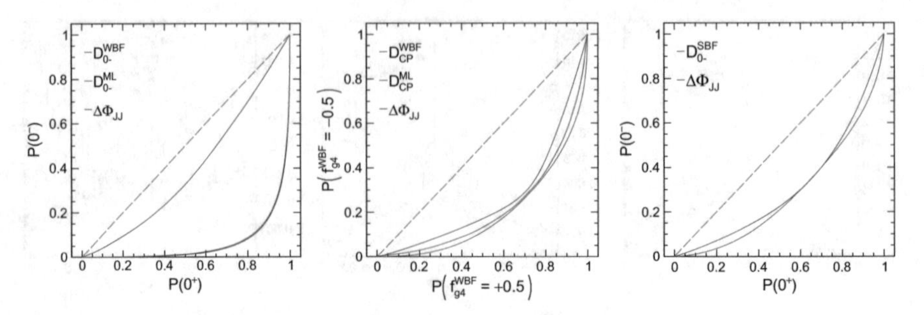

Fig. 4.14 Left: a ROC curve showing separation power of the scalar (SM-like 0^+) and pseudoscalar (0^-) models in the VBF process using the $\mathcal{D}_{0-}$ and $\Delta\Phi_{JJ}$ observables. The diagonal dashed line shows the hypothetical no-separation scenario. The points represent the efficiency of selecting each model as threshold in selecting the observable is varied. Right: same as left plot, but for the ggH process. with a requirement $m_{JJ} > 300$ GeV applied to enhance the VBF-like topology of events. Middle: a ROC curve showing separation power of the $f_{a3}^{\rm ggH} = +0.5$ and $f_{a3}^{\rm ggH} = -0.5$ models in the VBF process using the $\mathcal{D}_{CP}$ and $\Delta\Phi_{JJ}$ observables. Also shown on the left and middle plots are the ROC curves representing performance of the optimal observables obtained with machine learning techniques [30]

the ggH process. The gain in using the optimal observable in the ggH process is not as large as in VBF because of the smaller differences in dynamics of the scalar and pseudoscalar models, as both are generated by the higher-dimension operators with the same powers of q_i^2 in Eq. (4.1). While the $\mathcal{D}_{0-}$ observable incorporates all kinematic and dynamic information, the truly CP-sensitive observable $\mathcal{D}_{CP}$ does not rely on dynamics and provides the maximal separation between the models with the maximal mixing of the CP-even and CP-odd contributions and with the opposite phases. This suggests the optimal way to show the ROC curve as discrimination between the $f_{a3}^{\rm VBF} = \pm 0.5$ models, as shown for the VBF process in Fig. 4.14 (middle).

4.6.2.2 Machine Learning

The discriminants calculated with the matrix elements directly, as discussed in Sect. 4.6.2, are powerful tools in analysis of experimental data. Most importantly, they provide scientific insight into the problem under consideration. Nonetheless, there could be practical considerations which limit their application in certain cases. For example, events with partial reconstruction would require integration over unobserved degrees of freedom and substantial detector effects in reconstructed events with poor experimental resolution would require modeling of such effects with transfer functions. All of these effects can be taken into account, but may make calculations inefficient or impractical.

Machine learning is a popular approach to data analysis, especially with the growing computational power of computers. The problem of differentiating between the two models, as designed in Eq. (4.19), becomes a trivial task with supervised learning where two samples of events with the signal and alternative models are provided as input for training. One key aspect where the matrix element approach provides the insight is the set of input observables $\vec{\Omega}$. As long as a complete set of observables, sufficient for the matrix element calculations, is provided to the machine learning algorithm, the outcome of training is guaranteed to be a discriminant optimal for this task, that is equivalent to that in Eq. (4.19), as long as the proper training is applied. We illustrate this with such a discriminant $\mathcal{D}_{0-}^{\mathrm{ML}}$ in Fig. 4.14 (left) in application to the VBF process.

Application of the machine learning approach to the discriminant in Eq. (4.22) is not obvious, because it requires knowledge of quantum mechanics to isolate the interference component. Nonetheless, we provide the prescription how to obtain such a discriminant. The discriminant trained to differentiate the models with maximal quantum-mechanical mixing of the signal and alternative contributions and opposite phases becomes a machine-learning equivalent to that in Eq. (4.22), following the discussion in Sect. 4.6.2. The complete kinematic information of the event should be provided to the training. We illustrate this approach with such a discriminant $\mathcal{D}_{CP}^{\mathrm{ML}}$ in Fig. 4.14 (middle) in application to the VBF process.

To summarize, the matrix element technique, expressed in Eqs. (4.19) and (4.22), can be expanded with the help of machine learning with two important ingredients: the complete set of matrix-element input observables has to be used and the machine learning process should be based on the carefully prepared samples according to models discussed above. The machine learning approach is still based on the matrix element calculations, as the training samples are generated based on the same matrix elements as the discriminants in Eqs. (4.19) and (4.22) [30].

References

1. D. de Florian, C. Grojean, et al., Handbook of LHC Higgs cross sections: 4. deciphering the nature of the Higgs sector (2016). https://doi.org/10.23731/CYRM-2017-002. arXiv:1610.07922
2. A.M. Sirunyan, et al., Observation of Higgs boson decay to bottom quarks. Phys. Rev. Lett. **121**(12), 121801 (2018). https://doi.org/10.1103/PhysRevLett.121.121801. arXiv:1808.08242 [hep-ex]
3. M. Aaboud, et al., Observation of $H \rightarrow b\bar{b}$ decays and VH production with the ATLAS detector. Phys. Lett. **B786**, 59–86 (2018). https://doi.org/10.1016/j.physletb.2018.09.013. arXiv:1808.08238 [hep-ex]
4. A.M. Sirunyan, et al., Observation of the Higgs boson decay to a pair of τ leptons with the CMS detector. Phys. Lett. **B779**, 283–316 (2018). https://doi.org/10.1016/j.physletb.2018.02.004. arXiv:1708.00373 [hep-ex]
5. M. Aaboud, et al., Cross-section measurements of the Higgs boson decaying into a pair of τ-leptons in proton-proton collisions at $\sqrt{s} = 13$ TeV with the ATLAS detector. Phys. Rev. D **99**, 072001 (2019). https://doi.org/10.1103/PhysRevD.99.072001. arXiv:1811.08856 [hep-ex]
6. The LHC Higgs Cross Section Working Group, Handbook of LHC Higgs cross sections: 3. Higgs properties (2013). https://doi.org/10.5170/CERN-2013-004. arXiv:1307.1347
7. M. Tanabashi, et al., Review of particle physics. Phys. Rev. D **98**, 030001 (2018). https://doi.org/10.1103/PhysRevD.98.030001. [Online]. Available: https://link.aps.org/doi/10.1103/PhysRevD.98.030001

8. M. Tanabashi, et al., Review of particle physics 2019 update (2019). [Online]. Available: http://pdg.lbl.gov/2019/
9. A.M. Sirunyan, et al., Measurements of properties of the Higgs boson decaying into the four-lepton final state in pp collisions at $\sqrt{s} = 13$ TeV. JHEP **11**, 047 (2017). https://doi.org/10.1007/JHEP11(2017)047. arXiv:1706.09936 [hep-ex]
10. M. Aaboud, et al., Measurement of the Higgs boson mass in the $H \rightarrow ZZ* \rightarrow 4l$ and $H \rightarrow \gamma\gamma$ channels with $\sqrt{s} = 13$ TeV *pp* collisions using the AT-LAS detector. Phys. Lett. **B784**, 345–366 (2018). https://doi.org/10.1016/j.physletb.2018.07.050. arXiv:1806.00242 [hep-ex]
11. F. Caola, K. Melnikov, Constraining the higgs boson width with zz production at the lhc. Phys. Rev. D **88**, 054024 (2013). https://doi.org/10.1103/PhysRevD.88.054024. [Online]. Available: https://link.aps.org/doi/10.1103/PhysRevD.88.054024
12. S. Chatrchyan, et al., On the mass and spin-parity of the Higgs boson candidate via its decays to Z boson pairs. Phys. Rev. Lett. **110**, 081803 (2013). https://doi.org/10.1103/PhysRevLett.110.081803. arXiv:1212.6639 [hep-ex]
13. S. Chatrchyan, et al., Measurement of the properties of a Higgs boson in the four-lepton final state. Phys. Rev. D **89**, 092007 (2014). https://doi.org/10.1103/PhysRevD.89.092007. arXiv:1312.5353 [hep-ex]
14. V. Khachatryan, et al., Constraints on the spin-parity and anomalous HVV couplings of the Higgs boson in proton collisions at 7 and 8 TeV. Phys. Rev. D **92**, 012004 (2015). https://doi.org/10.1103/PhysRevD.92.012004. arXiv:1411.3441 [hep-ex]
15. V. Khachatryan, et al., Limits on the Higgs boson lifetime and width from its decay to four charged leptons. Phys. Rev. D **92**, 072010 (2015). https://doi.org/10.1103/PhysRevD.92.072010. arXiv:1507.06656 [hep-ex]
16. V. Khachatryan, et al., Combined search for anomalous pseudoscalar HVV couplings in VH($\mathrm{H} \rightarrow \mathrm{b\bar{b}}$) production and $\mathrm{H} \rightarrow$ VV decay. Phys. Lett. B **759**, 672 (2016). https://doi.org/10.1016/j.physletb.2016.06.004. arXiv:1602.04305 [hep-ex]
17. A.M. Sirunyan, et al., Constraints on anomalous Higgs boson couplings using production and decay information in the four-lepton final state. Phys. Lett. B **775**, 1 (2017). https://doi.org/10.1016/j.physletb.2017.10.021. arXiv:1707.00541 [hep-ex]
18. A.M. Sirunyan, et al., Measurements of the Higgs boson width and anomalous *HVV* couplings from on-shell and off-shell production in the four-lepton final state. Phys. Rev. D **99**(11), 112003 (2019). https://doi.org/10.1103/PhysRevD.99.112003. arXiv:1901.00174 [hep-ex]
19. A.M. Sirunyan, et al., Constraints on anomalous *HVV* couplings from the production of Higgs bosons decaying to τ lepton pairs. Phys. Rev. D **100**(11), 112002 (2019). https://doi.org/10.1103/PhysRevD.100.112002. arXiv:1903.06973 [hep-ex]
20. CMS collaboration, Constraints on anomalous Higgs boson couplings to vector bosons and fermions in production and decay $H \rightarrow 4l$channel. Report No.: CMS-PAS-HIG-19-009, to be submitted to *Phys. Rev. D*. http://cds.cern.ch/record/2725543
21. G. Aad, et al., Evidence for the spin-0 nature of the Higgs boson using ATLAS data. Phys. Lett. B **726**, 120 (2013). https://doi.org/10.1016/j.physletb.2013.08.026. arXiv:1307.1432 [hep-ex]
22. G. Aad, et al., Study of the spin and parity of the Higgs boson in diboson decays with the ATLAS detector. Eur. Phys. J. C **75**, 476 (2015). https://doi.org/10.1140/epjc/s10052-015-3685-1. arXiv:1506.05669 [hep-ex]
23. G. Aad, et al., Test of CP invariance in vector-boson fusion production of the Higgs boson using the optimal observable method in the ditau decay channel with the ATLAS detector. Eur. Phys. J. C **76**, 658 (2016). https://doi.org/10.1140/epjc/s10052-016-4499-5. arXiv:1602.04516 [hep-ex]
24. G. Aad, et al., Constraints on the off-shell Higgs boson signal strength in the high-mass ZZ and WW final states with the ATLAS detector. Eur. Phys. J. C **75**, 335 (2015). https://doi.org/10.1140/epjc/s10052-015-3542-2. arXiv:1503.01060 [hep-ex]
25. M. Aaboud, et al., Measurement of inclusive and differential cross sections in the $H \rightarrow ZZ* \rightarrow 4l$ decay channel in *pp* collisions at $\sqrt{s} = 13$ TeV with the ATLAS detector. J. High Energy Phys. **2017**(10), 132 (2017). ISSN: 1029-8479. https://doi.org/10.1007/JHEP10(2017)132

26. The ATLAS Collaboration, et al., Measurement of the Higgs boson coupling properties in the $H \rightarrow ZZ* \rightarrow 4l$ decay channel at sqrts= 13 TeV with the ATLAS detector. J. High Energy Phys. **2018**(3), 95 (2018). https://doi.org/10.1007/JHEP03(2018)095
27. M. Aaboud, et al., Measurements of Higgs boson properties in the diphoton decay channel with 36 fb^{-1} of pp collision data at $\sqrt{s} = 13$ TeV with the ATLAS detector. Phys. Rev. D **98**, 052005 (2018). https://doi.org/10.1103/PhysRevD.98.052005. [Online]. Available: https://link.aps.org/doi/10.1103/PhysRevD.98.052005
28. I. Anderson, S. Bolognesi, F. Caola, Y. Gao, A.V. Gritsan, C.B. Martin, K. Melnikov, M. Schulze, N.V. Tran, A. Whitbeck, Y. Zhou, Constraining anomalous HVV interactions at proton and lepton colliders. Phys. Rev. D **89**, 035007 (2014). https://doi.org/10.1103/PhysRevD.89.035007. arXiv:1309.4819 [hep-ph]
29. A.V. Gritsan, R. Röntsch, M. Schulze, M. Xiao, Constraining anomalous Higgs boson couplings to the heavy flavor fermions using matrix element techniques. Phys. Rev. D **94**, 055023 (2016). https://doi.org/10.1103/PhysRevD.94.055023. arXiv:1606.03107 [hep-ph]
30. A. V. Gritsan, J. Roskes, U. Sarica, M. Schulze, M. Xiao, and Y. Zhou, "New features in the JHU generator framework: constraining Higgs boson properties from on-shell and off-shell production," *Phys. Rev. D*, vol. 102, no. 5, p. 056022, 2020. https://doi.org/10.1103/PhysRevD.102.056022.arXiv:2002.09888 [hep-ph]
31. Y. Gao, A.V. Gritsan, Z. Guo, K. Melnikov, M. Schulze, N.V. Tran, Spin determination of single-produced resonances at hadron colliders. Phys. Rev. D **81**, 075022 (2010). https://doi.org/10.1103/PhysRevD.81.075022. arXiv:1001.3396 [hep-ph]
32. S. Bolognesi, Y. Gao, A.V. Gritsan, K. Melnikov, M. Schulze, N.V. Tran, A. Whitbeck, Spin and parity of a single-produced resonance at the LHC. Phys. Rev. D **86**, 095031 (2012). https://doi.org/10.1103/PhysRevD.86.095031. arXiv:1208.4018 [hep-ph]
33. E. Boos, M. Dobbs, et al., Generic user process interface for event generators (2001). arXiv:hep-ph/0109068
34. J. Alwall, A. Ballestrero, et al., A standard format for Les Houches event files (2006). https://doi.org/10.1016/j.cpc.2006.11.010. arXiv:hep-ph/0609017
35. J.M. Campbell, R.K. Ellis, MCFM for the Tevatron and the LHC. Nucl. Phys. Proc. Suppl. **205-206**, 10 (2010). https://doi.org/10.1016/j.nuclphysbps.2010.08.011. arXiv:1007.3492 [hep-ph]
36. S. Frixione, P. Nason, C. Oleari, Matching NLO QCD computations with parton shower simulations: the POWHEG method. JHEP **11**, 070 (2007). https://doi.org/10.1088/1126-6708/2007/11/070. arXiv:0709.2092 [hep-ph]
37. E. Bagnaschi, G. Degrassi, P. Slavich, A. Vicini, Higgs production via gluon fusion in the POWHEG approach in the SM and in the MSSM. JHEP **02**, 088 (2012). https://doi.org/10.1007/JHEP02(2012)088. arXiv:1111.2854 [hep-ph]
38. P. Nason, C. Oleari, NLO Higgs boson production via vector-boson fusion matched with shower in POWHEG. JHEP **02**, 037 (2010). https://doi.org/10.1007/JHEP02(2010)037. arXiv:0911.5299 [hep-ph]
39. H.B. Hartanto, B. Jager, L. Reina, D.Wackeroth, Higgs boson production in association with top quarks in the POWHEG BOX. Phys. Rev. D **91**, 094003 (2015). https://doi.org/10.1103/PhysRevD.91.094003. arXiv:1501.04498 [hep-ph]
40. K. Hamilton, P. Nason, G. Zanderighi, MINLO: multi-scale improved NLO. JHEP **10**, 155 (2012). https://doi.org/10.1007/JHEP10(2012)155. arXiv:1206.3572 [hep-ph]
41. G. Luisoni, P. Nason, C. Oleari, F. Tramontano, HW$^{\pm}$/HZ + 0 and 1 jet at NLO with the POWHEG BOX interfaced to GoSam and their merging within MiNLO. JHEP **10**, 083 (2013). https://doi.org/10.1007/JHEP10(2013)083. arXiv:1306.2542 [hep-ph]
42. J. Neyman, E.S. Pearson, On the problem of the most efficient tests of statistical hypotheses. Phil. Trans. R. Soc. **A231**, 289–337 (1933). https://doi.org/10.1098/rsta.1933.0009
43. T. Plehn, D.L. Rainwater, D. Zeppenfeld, Determining the structure of Higgs couplings at the LHC. Phys. Rev. Lett. **88**, 051801 (2002). https://doi.org/10.1103/PhysRevLett.88.051801. arXiv:hep-ph/0105325 [hep-ph]

Chapter 5
Higgs Boson Data Analysis

With the data processed and recorded and the theoretical basis for the calculations understood, we can analyze the data to measure the Higgs boson's properties. Figs. 5.1 and 5.2 shows the mass distribution of the four-lepton events from the Run 2 dataset, with the Higgs boson peak at 125 GeV shown in red. Most of the analyses shown here use some or all of these events.

5.1 Run 1 Results

Measuring the spin and parity of the Higgs boson was one of the first experimental priorities after its discovery. The earliest papers from Run 1 of the LHC confirmed that the newly discovered particle primarily interacts as a spin-zero particle with $J^{CP} = 0^{++}$, with results from both CMS [5–7] and ATLAS [8, 9]. The spin analyses used the Higgs boson's decay to $\mathrm{H} \to \mathrm{ZZ} \to 4\ell$, $\mathrm{H} \to \mathrm{WW} \to 2\ell 2\nu$, and $\gamma\gamma$, while the parity analyses, which need more degrees of freedom than a two-body decay can provide, used $\mathrm{H} \to \mathrm{ZZ} \to 4\ell$ and $\mathrm{H} \to \mathrm{WW} \to 2\ell 2\nu$.

As a validation of the spin analysis, it is also interesting to measure the spin of the Z boson, which is well known to be a spin-1 particle, using identical methods. This serves as a validation of the matrix element procedure as well as of the background modeling. The Z boson can decay to 4 leptons through the diagram shown in Fig. 5.3, which forms the peak at 91.2 GeV in Fig. 5.2. This is distinguished from the alternate hypothesis of a new H boson at 91.2 GeV that decays to 4 leptons via $\mathrm{H} \to \mathrm{ZZ} \to 4\ell$. It is also possible that the Z boson exists and behaves as expected, but a tiny fraction of the peak f_{H} is made by a new Higgs boson. Using methods similar to the ones that will be described below, the fractional contribution of a spin 0 particle to the Z peak is measured to be less than 0.8% at 95% confidence level, as shown in Fig. 5.4.

J. Roskes, *A Boson Learned from its Context, and a Boson Learned from its End*, Springer Theses, https://doi.org/10.1007/978-3-030-58011-7_5

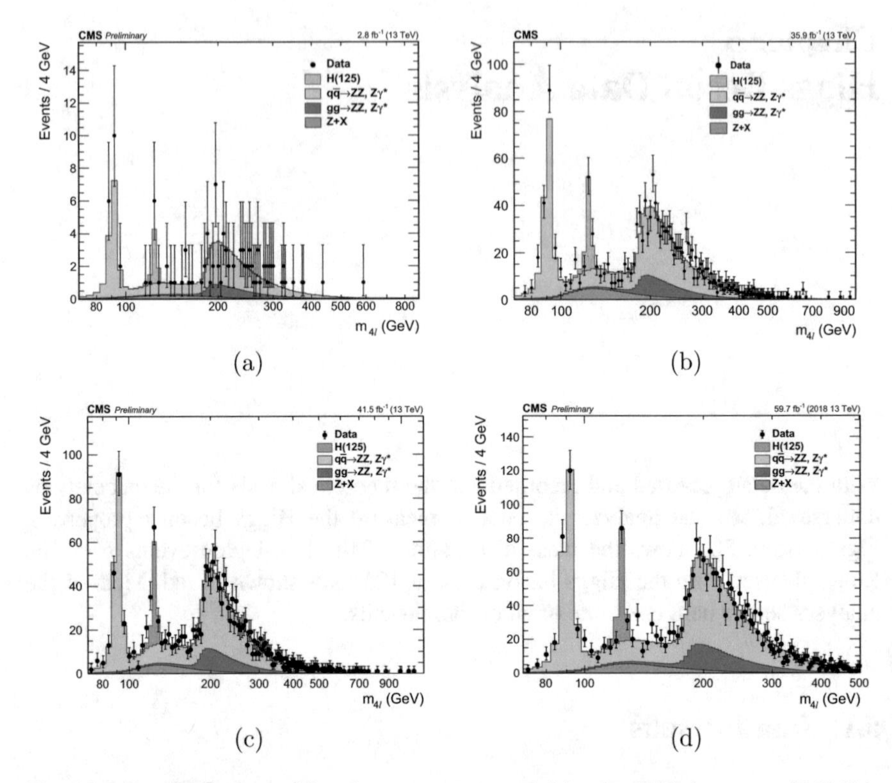

Fig. 5.1 Mass distributions of four-lepton events recorded by the CMS detector at 13 TeV in (**a**) 2015 [1], (**b**) 2016 [2], (**c**) 2017 [3], and (**d**) 2018 [4]

After hypotheses such as a pure a_3 contribution were excluded, it still remains interesting to search for a small anomalous contribution to the Higgs boson's interactions. The first comprehensive paper, searching for a wide variety of alternate spin and coupling hypotheses, was [7], which used $\mathrm{H} \to \mathrm{ZZ} \to 4\ell$, $\mathrm{H} \to \mathrm{WW} \to 2\ell 2\nu$, and $\mathrm{H} \to \gamma\gamma$ data. These early analyses form the starting point for the more complicated analyses in Run 2, to be discussed further.

The simplest analyses assumed that a maximum of one anomalous term is nonzero and that the anomalous couplings are real, so that the amplitude and probability for the Higgs boson's decay to four fermions, as a function of the SM coupling a_1, an anomalous coupling a_i, and the lepton kinematics $\vec{\Omega}$ is

$$\mathcal{A}(a_1, a_i, \vec{\Omega}) = a_1 \mathcal{A}_1(\vec{\Omega}) + a_i \mathcal{A}_i(\vec{\Omega}) \tag{5.1}$$

$$\mathcal{P}(a_1, a_i, \vec{\Omega}) = |\mathcal{A}|^2 = a_1^2 \mathcal{P}_1(\vec{\Omega}) + a_i^2 \mathcal{P}_i(\vec{\Omega}) + a_1 a_i \mathcal{P}_{\text{int}}(\vec{\Omega}) \tag{5.2}$$

The analysis proceeds by constructing *templates*, n-dimensional histograms that parameterize the probability as a function of Ω, for $\mathcal{P}_1$, $\mathcal{P}_i$, and $\mathcal{P}_{\text{int}}$, as well as

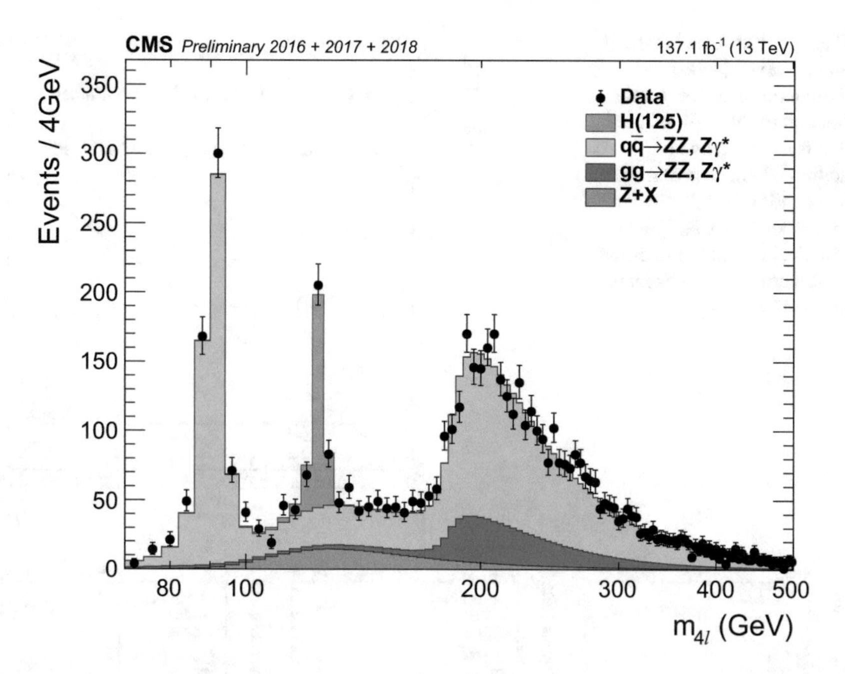

Fig. 5.2 Mass distribution of all of the four-lepton events recorded by the CMS detector in 2016, 2017, and 2018 [4]

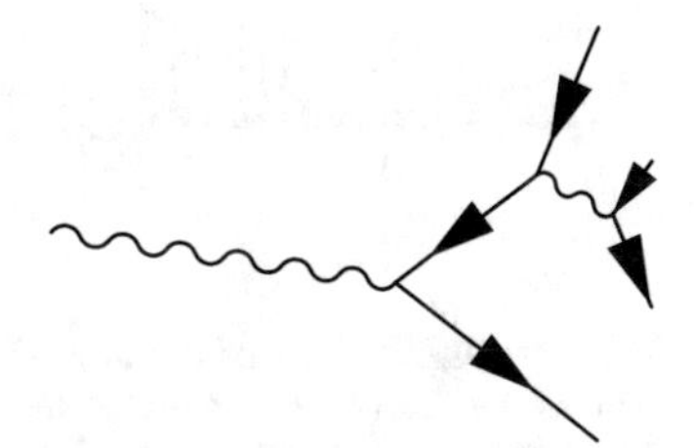

Fig. 5.3 Feynman diagram for the Z boson's decay to 4 leptons

for the background contributions. The signal templates are all constructed from gluon fusion Monte Carlo produced by POWHEG [10–14] with the $\mathrm{H} \to \mathrm{ZZ} \to 4\ell$ decay provided by JHUGEN [15–19]. The irreducible backgrounds for this analysis are $\mathrm{qq} \to 4\ell$, also estimated through Monte Carlo simulated by POWHEG, and $\mathrm{gg} \to 4\ell$, estimated through MCFM [20] simulation. Additionally, the $\mathrm{Z} + \mathrm{X}$ background, which comes primarily from jets that are misinterpreted as leptons in the detector, is estimated using a control region in the data.

In principle, the ideal way to go would be to construct templates using the full probability distribution as a function of the angles and masses that define $\vec{\Omega}$, as shown in Fig. 4.8. This was done in some simplified cases in that paper, but does not scale well. Instead, we project $\vec{\Omega}$ onto the most relevant degrees of freedom using the MELA discriminants described in Sect. 4.6.2 and bin the distribution in 3D histograms. For an analysis that searches for just one anomalous coupling, it is possible to choose three observables that lose no information: $\mathcal{D}_{\mathrm{bkg}}$, which separates

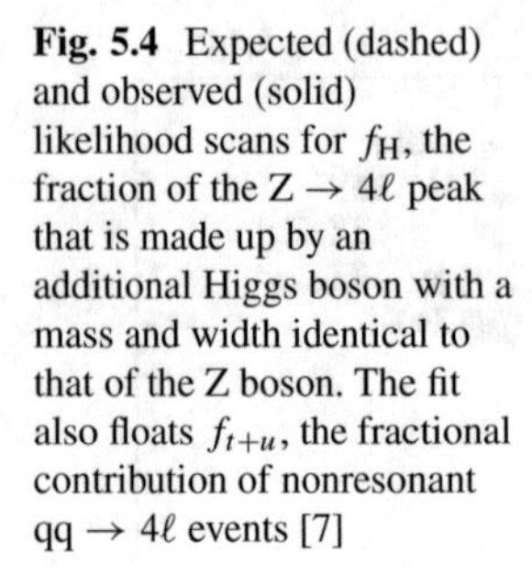

Fig. 5.4 Expected (dashed) and observed (solid) likelihood scans for f_{H}, the fraction of the $\mathrm{Z} \rightarrow 4\ell$ peak that is made up by an additional Higgs boson with a mass and width identical to that of the Z boson. The fit also floats f_{t+u}, the fractional contribution of nonresonant $\mathrm{qq} \rightarrow 4\ell$ events [7]

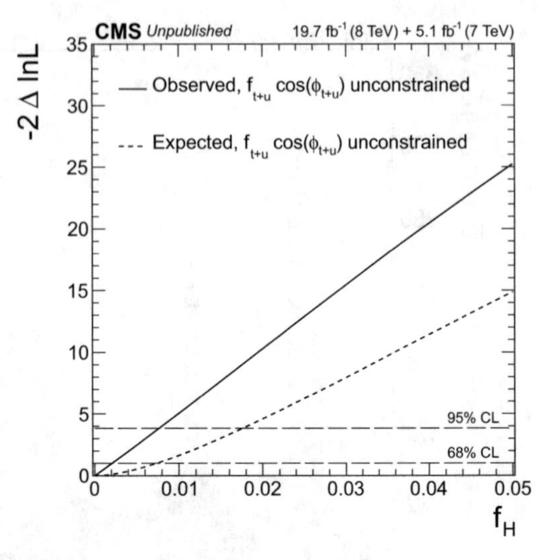

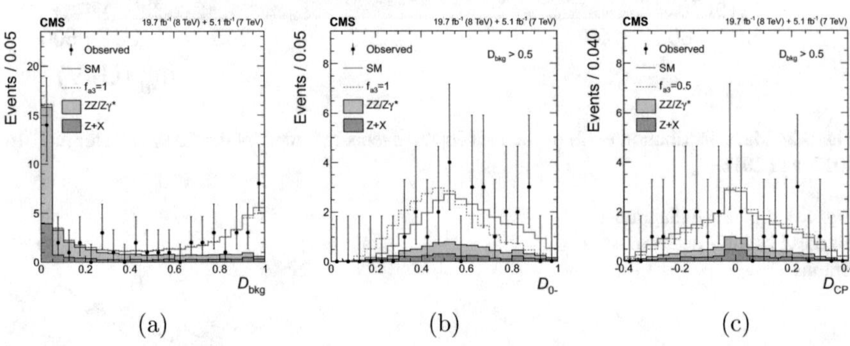

Fig. 5.5 Distributions of the three discriminants for the Run 1 measurement of f_{a3}. (**a**) $\mathcal{D}_{\mathrm{bkg}}$, defined by Eq. (4.19), separates the SM signal from background. (**b**) $\mathcal{D}_{0-}$, also defined by Eq. (4.19), separates SM signal from pure a_3 signal. (**c**) $\mathcal{D}'_{CP}$, defined by Eq. (4.23), separates the interference component between a_1 and a_3 [7]

signal from background; a $\mathcal{D}_{ai}$ discriminant to separate the SM coupling from the chosen BSM coupling a_i; and a $\mathcal{D}_{\mathrm{int}}$ discriminant to separate the interference contribution. $\mathcal{D}_{\mathrm{bkg}}$ is calculated from the reconstructed Higgs boson's invariant mass $m_{4\ell}$ as well as the kinematics from the decay, the angles and dilepton invariant masses in Fig. 4.8. The other discriminants rely only on the decay kinematics. Figure 5.5 shows the distributions of some of these discriminants in the Monte Carlo simulation and data.

The interference discriminant shown in Fig. 5.5c is special in the sense that it represents interference between a CP-even and a CP-odd process. The distribution of this discriminant for any purely CP-even (such as a pure SM Higgs boson or any background process) or CP-odd process (such as a pure a_3 Higgs boson) will be symmetric around 0, as shown in the figure. Although this analysis and other similar ones described below search for nonzero f_{a3}, any statistically significant asymmetry in $\mathcal{D}_{CP}$ would be a sign of CP violation, even if it does not match a particular hypothesis. Another interference discriminant is used in the analysis measuring a_2, which detects the interference between a_1 and a_2, but that discriminant shows no special symmetry because the interference is between two CP-even terms.

In this simplest example, the *only* contributions to the probability are background, SM signal, pure BSM signal, and interference. For this reason, three discriminants are sufficient to contain all the information from the kinematics, as described in Sect. 4.6.2. A small amount of information is lost due to finite binning of those discriminants, but enough bins were used that the loss is small.

Once the templates are constructed, we perform an unbinned extended maximum likelihood fit [21], where the probability density is normalized to the total event yield in each process j and category k. In the analyses here, the events were divided into categories depending on the final state lepton flavor: $\mathrm{H} \rightarrow 2\mathrm{e}2\mu$, 4e, or 4μ, and the signal processes are all included together, but the notation is general to accommodate later, more complicated analyses. The overall probability density function is given by

$$\mathcal{P}_{jk}(\vec{\Omega}; \mu_j, \vec{f}_j) = \mu_j \mathcal{P}^{\text{sig}}_{jk}\left(\vec{\Omega}; \vec{f}\right) + \mathcal{P}^{\text{bkg}}_{jk}\left(\vec{\Omega}\right), \tag{5.3}$$

$\mathcal{P}^{\text{sig}}$ is defined by Eq. (5.2) for these analyses and similar expressions for the more complicated ones described later. It is a function of the kinematics $\vec{\Omega}$, the anomalous couplings $\vec{f}$, and the overall scaling μ. As described in Sect. 4.4, we reparameterize the SM coupling a_1 and n anomalous couplings $\vec{a}_i$ into n f_{ai}'s (Eq. (4.2)), one for each anomalous coupling, and the signal strength μ. In this way, we decorrelate the shape of the event distributions, which is our primary interest in these analyses, from the number of events. In more complicated analyses, different signal processes will have separate μ_j's.

The result of the analysis is a likelihood scan that gives the log likelihood for each value of f_{ai}. At each point in the scan, μ as well as various systematic uncertainties are profiled, so that the result is independent of the signal yield. Any value of f_{ai} where the log likelihood is above the lower dotted line is excluded at 68% confidence level, and any point above the upper dotted line is excluded at 95% confidence level. Some of the scans from Run 1 are shown in Fig. 5.6.

The same paper also included fits for two simultaneous anomalous couplings, with an amplitude and probability distribution given by

$$\mathcal{A}(a_1, a_i, a_j, \vec{\Omega}) = a_1 \mathcal{A}_1(\vec{\Omega}) + a_i \mathcal{A}_i(\vec{\Omega}) + a_j \mathcal{A}_j(\vec{\Omega}) \tag{5.4}$$

$$\begin{aligned}\mathcal{P}(a_1, a_i, a_j, \vec{\Omega}) = |\mathcal{A}|^2 =& a_1^2 \mathcal{P}_1(\vec{\Omega}) + a_i^2 \mathcal{P}_i(\vec{\Omega}) + a_j^2 \mathcal{P}_j(\vec{\Omega}) \\ &+ a_1 a_i \mathcal{P}^{1i}_{\text{int}}(\vec{\Omega}) + a_1 a_j \mathcal{P}^{1j}_{\text{int}}(\vec{\Omega}) + a_i a_j \mathcal{P}^{ij}_{\text{int}}(\vec{\Omega})\end{aligned} \tag{5.5}$$

Note that the number of signal terms has increased from 3 to 6, and there is no longer a way to provide optimal separation between all the terms with only 3 discriminants. These results are not shown here, but this equation shows how the number of

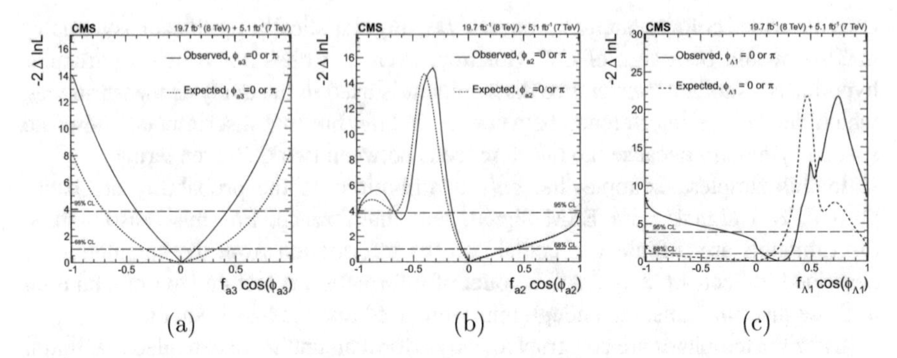

Fig. 5.6 Expected (dashed) and observed (solid) likelihood scans for the effective fractions (**a**) f_{a3}, (**b**) f_{a2}, and (**c**) $f_{\Lambda 1}$ from Run 1 $\mathrm{H} \to \mathrm{ZZ} \to 4\ell$ events with all other anomalous couplings fixed to 0. The $\cos\phi_{ai}$ term allows a signed quantity, where $\cos\phi_{ai} = -1$ or $+1$ [7]

terms grows with more couplings, which will be revisited later. Additionally, some analyses allowed the couplings to be complex, which requires a second interference term for each pair of couplings.

Later analyses used Run 1 data to search for anomalous couplings in production: VH in the case of CMS [22] and VBF in the case of ATLAS [23]. As mentioned in Sect. 4.4, production is sensitive to small anomalous couplings; however, due to the low statistics available from Run 1 data, the results were at a low confidence level.

In addition, CMS [24] and ATLAS [25] searched for offshell Higgs boson production and put constraints on its width. The CMS analysis searched for the Λ_Q coupling from Eq. (4.1), and these results are currently the only constraints on this coupling.

5.2 First Run 2 Results

The first CMS $\mathrm{H} \to \mathrm{ZZ} \to 4\ell$ analysis in Run 2 [1] used the data taken in 2015. Using the first year of 13 TeV data, CMS observed the Higgs boson peak in Fig. 5.1a at a confidence level of 2.5σ. The analysis also searched for events produced in vector boson fusion.

With the increased luminosity in 2016, the data, shown in Fig. 5.1b, were sufficient to conduct more detailed analyses of the Higgs boson's properties [2], including the first anomalous coupling measurements in production (the Higgs boson's "context") and decay (its "end") at the same time [26], using kinematics of VBF and VH production, where the associated vector boson in VH production decays to quarks. The results shown here are from the next iteration of this analysis [27], which used the same strategies applied to the data from 2016 and 2017.

In order to isolate VBF and VH, a MELA discriminant $\mathcal{D}_{\mathrm{2jet}}$ is used to separate VBF and VH production from gluon fusion produced in association with two jets.

This discriminant is defined by Eq. (4.19), with VBF or VH production in the numerator and gluon fusion in the denominator. For VBF or VH production, each analysis uses the maximum of the probability under the SM and the probability under the pure anomalous hypothesis considered in that analysis. In this way, the categorization efficiently selects both SM and BSM events, for greater sensitivity. Other requirements on the number of jets and leptons in the event are also applied in order to suppress the $t\bar{t}$H contribution.

- The VBF-tagged category requires exactly four leptons, either two or three jets of which at most one is b-quark flavor-tagged, or more if none are b-tagged jets, and $\mathcal{D}_{\text{2jet}}^{\text{VBF}} > 0.5$ using either the SM or the BSM signal hypotheses for the VBF production.
- The VH-tagged category requires exactly four leptons, either two or three jets, or more if none are b-tagged jets, and $\mathcal{D}_{\text{2jet}}^{\text{VH}} = \max\left(\mathcal{D}_{\text{2jet}}^{\text{WH}}, \mathcal{D}_{\text{2jet}}^{\text{ZH}}\right) > 0.5$ using either the SM or the BSM signal hypothesis for the VH production.
- The untagged category contains the remaining events.

Plots of the MELA discriminants used for categorization are shown in Fig. 5.7, using the f_{a3} analysis as an example.

For VBF or VH production of a Higgs boson that subsequently decays H $\to$ ZZ $\to 4\ell$, the HVV vertex appears twice: once on the production side and once on the decay side. Equation (5.2) is modified to:

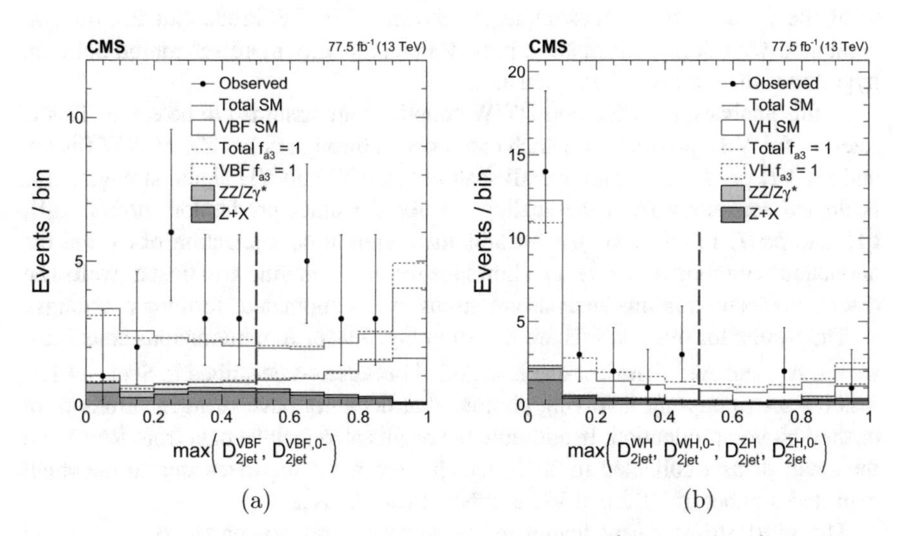

Fig. 5.7 Distributions of events for the discriminants $\max\left(\mathcal{D}_{\text{2jet}}^{\text{VBF}}, \mathcal{D}_{\text{2jet}}^{\text{VBF},0-}\right)$ (left) and $\max\left(\mathcal{D}_{\text{2jet}}^{\text{WH}}, \mathcal{D}_{\text{2jet}}^{\text{WH},0-}, \mathcal{D}_{\text{2jet}}^{\text{ZH}}, \mathcal{D}_{\text{2jet}}^{\text{ZH},0-}\right)$ (right) from the analysis of the a_3 coupling for a pseudoscalar contribution. The requirement $\mathcal{D}_{\text{bkg}} > 0.5$ is applied in order to enhance the signal contribution over the background. The VBF signal under both the SM and pseudoscalar hypotheses is enhanced in the region above 0.5 for the former variable, and the WH and ZH signals are similarly enhanced in the region above 0.5 for the latter variable [27]

$$\mathcal{A}(a_1, a_i, \vec{\Omega}) = \left(a_1 \mathcal{A}_1^{\text{prod}}(\vec{\Omega}) + a_i \mathcal{A}_i^{\text{prod}}(\vec{\Omega})\right)\left(a_1 \mathcal{A}_1^{\text{dec}}(\vec{\Omega}) + a_i \mathcal{A}_i^{\text{dec}}(\vec{\Omega})\right) \tag{5.6}$$

$$\mathcal{P}(a_1, a_i, \vec{\Omega}) = |\mathcal{A}|^2 = a_1^4 \mathcal{P}_0(\vec{\Omega}) + a_1^3 a_i \mathcal{P}_1(\vec{\Omega}) + a_1^2 a_i^2 \mathcal{P}_2(\vec{\Omega}) + a_1 a_i^3 \mathcal{P}_3(\vec{\Omega}) + a_i^4 \mathcal{P}_4(\vec{\Omega}) \tag{5.7}$$

There are now five terms in the probability, each represented by a template, which is constructed using Monte Carlo simulation from JHUGEN. The gluon fusion contribution to the probability is unchanged and still follows Eq. (5.2), with templates modeled through POWHEG+JHUGEN simulations.

In each category, discriminants are chosen to best utilize the information provided by the production mode targeted by that category. In the untagged category, which does not target any specific production mode and is dominated by gluon fusion, the same setup as in Run 1 is used. In the VBF and VH categories, in principle, we would need four discriminants to separate between the five terms in Eq. (5.7), plus another one to separate signal from background. Using so many discriminants is impractical, so we choose the ones that separate the most useful degrees of freedom: $\mathcal{D}_{\text{bkg}}^{\text{VBF/VH+dec}}$, which separates signal from background using the product of the production and decay probabilities; $\mathcal{D}_{\text{alt}}^{\text{VBF/VH+dec}}$, which separates SM signal from pure BSM signal, again using both the production and decay probabilities; and $\mathcal{D}_{\text{int}}^{\text{VBF/VH}}$, which separates the interference component for production. In the VBF-tagged category, VBF information is used, while in the VH category, VH information is used. Distributions of these discriminants, again using the f_{a3} analysis as an example, are shown in Fig. 5.8. Production information combined with decay information provides significantly more separation between hypotheses than decay information alone.

In this analysis, the HZZ and HWW couplings are assumed to be equal. This is relevant for VBF production, which can proceed through either ZZ or WW fusion, and for VH production. The overall scaling for VBF and VH signal strength, μ_V, is floated separately from the scaling μ_F for the other production modes, ggH, $t\bar{t}$H, and bbH. In this way, μ_V absorbs the common normalization of a_1 and the anomalous coupling a_i, while μ_F allows the fermion coupling κ to float as well. The discriminants used in this analysis are insensitive to anomalous fermion couplings.

The results for this analysis are shown in Fig. 5.9 for four anomalous couplings: a_3, a_2, Λ_1, and $\Lambda_1^{Z\gamma}$. The last one is included because, as described in Sect. 4.4.1.1, it is the only coupling involving photons that does not have stringent limits from onshell photon production. In addition, the results also include data from Run 1 and the small dataset collected in 2015, which were not categorized due to the small expected number of VBF and VH events in those datasets.

The most striking new feature of these results, as compared to the ones in Fig. 5.6, is a narrow but shallow minimum around $f_{ai} = 0$. This is a result of the discussion in Sect. 4.4.1. Because the anomalous couplings are proportional to q^2 of the vector bosons, and because the vector bosons in VBF and VH production have a higher q^2, VBF and VH are sensitive to smaller anomalous couplings than decay is. Conversely, when the anomalous couplings are *large* ($f_{a3} \gtrsim 0.005$, with slightly

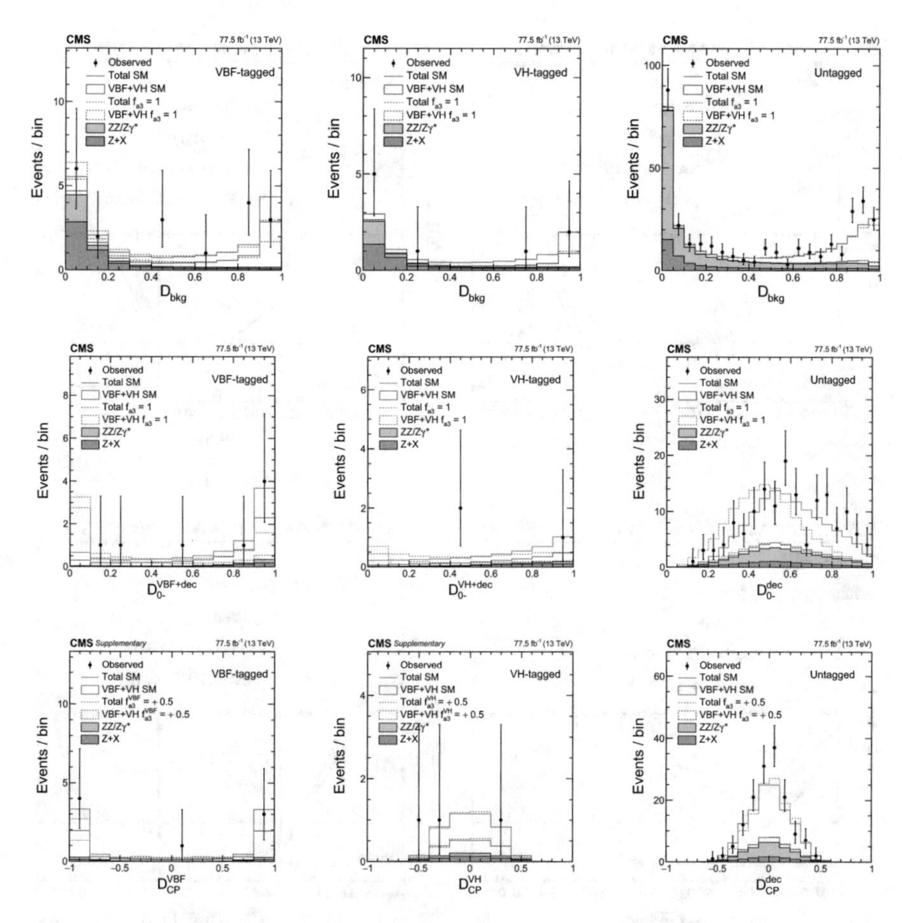

Fig. 5.8 Distributions of the three discriminants used to measure f_{a3} in the three categories. The top row shows $\mathcal{D}_{\text{bkg}}$ (Eq. (4.19)), the middle row shows $\mathcal{D}_{0-}$ (also Eq. (4.19)), and the bottom row shows $\mathcal{D}_{CP}$ (Eq. (4.22)). The type of kinematic information used for each discriminant depends on the category. The left column shows the discriminants in the VBF-tagged category, the middle column shows the ones in the VH-tagged category, and the right column shows the discriminants for the untagged category, which can be compared to Fig. 5.5. All of the plots except $\mathcal{D}_{\text{bkg}}$ use a requirement $\mathcal{D}_{\text{bkg}} > 0.5$ in order to enhance the signal over background contributions [27]

different numerical values for the other anomalous couplings), the SM contribution to VBF and VH is much smaller than the anomalous contribution, and further increases in f_{a3} do not affect the VBF shape. The minimum is shallow because its depth is limited by the relatively small number of events in the VBF- and VH-tagged categories, which can be seen by counting events in Fig. 5.8.

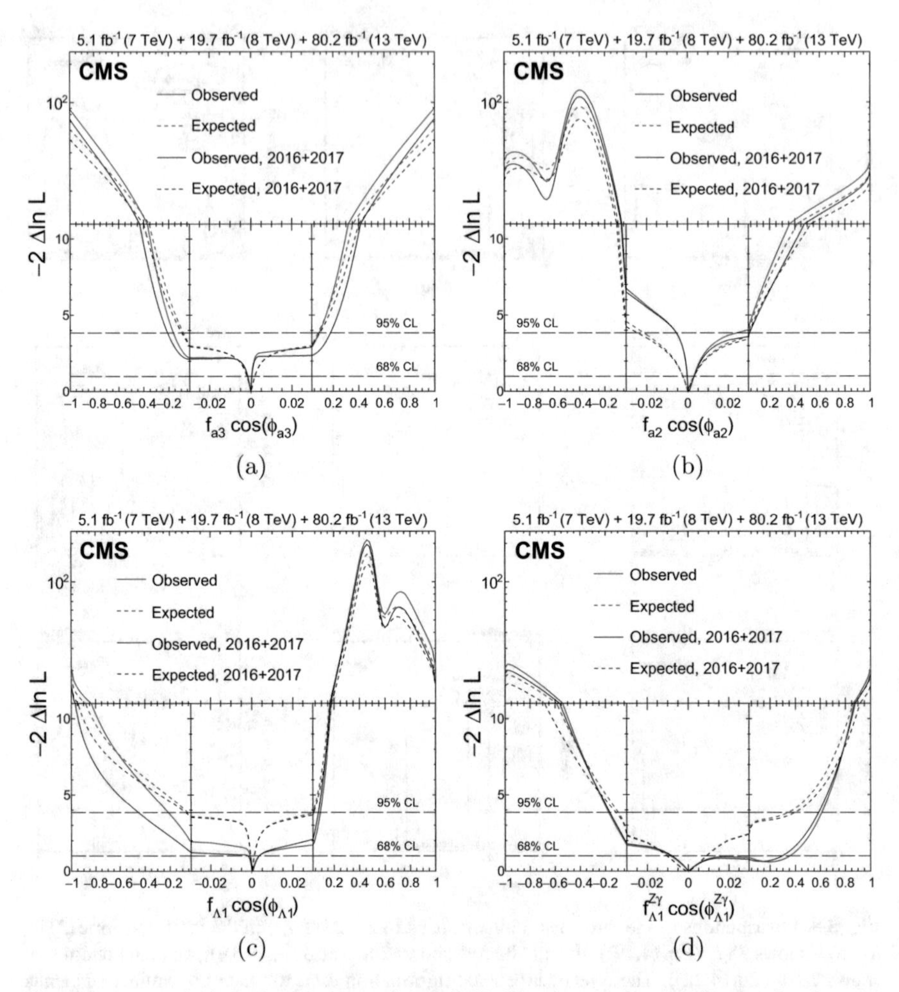

Fig. 5.9 Expected (dashed) and observed (solid) likelihood scans for the effective fractions (**a**) f_{a3}, (**b**) f_{a2}, (**c**) $f_{\Lambda 1}$, and (**d**) $f_{\Lambda 1}^{Z\gamma}$ from VBF and VH production and $\mathrm{H} \rightarrow \mathrm{VV} \rightarrow 4\ell$ decay information from four-lepton events, with all other anomalous couplings fixed to 0 [27]

5.3 Offshell Higgs Boson Properties

The same paper [27] also includes measurements of anomalous couplings in the offshell region above 200 GeV. Similar to VBF and VH production, the offshell region is sensitive to smaller anomalous couplings than the onshell region, because both Z bosons from the decay are onshell, with a mass of around 91.2 GeV. By contrast, in the onshell region the lighter Z boson often has an invariant mass around 30 GeV, as shown in Fig. 4.4a. In this way, the offshell region can provide additional sensitivity to anomalous couplings. The left plot of Fig. 4.6 shows this effect: anomalous couplings result in an increased number of events in the offshell region.

Another interesting parameter, which is correlated with anomalous couplings, is the Higgs boson's width. As described in Sect. 4.3, the cross section to produce an offshell Higgs boson is proportional to its width. Seeing a higher-than-expected number of events in the offshell region can be a result of either a larger width or anomalous couplings. To distinguish between them, we use the same kinds of MELA discriminants as in the onshell region. Afterwards, we scan the anomalous couplings and float the width, and separately scan the width and float anomalous couplings. In this way, the measurement uses as few assumptions as possible.

The offshell measurement is more complicated because of interference between signal and background. (In the onshell case, this interference is essentially zero because of the narrow peak at 125 GeV, and we neglect it in all onshell analyses.) Each process in the offshell region interferes with background processes with the same initial and final states and similar topology. Gluon fusion interferes with gg → ZZ; VBF interferes with vector boson scattering, which is the same Feynman diagram as VBF but involving vertices of three or four Z or W bosons and no Higgs boson; and VH interferes with VVV production. The result is that the signal probability density function for gluon fusion (5.2) becomes

$$\mathcal{A}(a_1, a_i, \vec{\Omega}) = a_1\mathcal{A}_1(\vec{\Omega}) + a_i\mathcal{A}_i(\vec{\Omega}) + \mathcal{A}_{\text{bkg}}(\vec{\Omega}) \tag{5.8}$$

$$\begin{aligned}\mathcal{P}(a_1, a_i, \vec{\Omega}) = |\mathcal{A}|^2 =& a_1^2\mathcal{P}_1(\vec{\Omega}) + a_i^2\mathcal{P}_i(\vec{\Omega}) + a_1a_i\mathcal{P}_{\text{int}}^{1i}(\vec{\Omega}) \\ &+a_1\mathcal{P}_{\text{bkgint}}^{1}(\vec{\Omega}) + a_i\mathcal{P}_{\text{bkgint}}^{i}(\vec{\Omega}) \\ +& \mathcal{P}_{\text{bkg}}(\vec{\Omega})\end{aligned} \tag{5.9}$$

The first and last lines of Eq. (5.9) were already included in Eqs. (5.2) and (5.3), while the middle line is new. Similarly, the VBF and VH probability Eq. (5.7) becomes

$$\begin{aligned}\mathcal{A}(a_1, a_i, \vec{\Omega}) =& \left(a_1\mathcal{A}_1^{\text{prod}}(\vec{\Omega}) + a_i\mathcal{A}_i^{\text{prod}}(\vec{\Omega})\right)\left(a_1\mathcal{A}_1^{\text{dec}}(\vec{\Omega}) + a_i\mathcal{A}_i^{\text{dec}}(\vec{\Omega})\right) \\ +& \mathcal{A}_{\text{bkg}}(\vec{\Omega})\end{aligned} \tag{5.10}$$

$$\begin{aligned}\mathcal{P}(a_1, a_i, \vec{\Omega}) = |\mathcal{A}|^2 =& a_1^4\mathcal{P}_0(\vec{\Omega}) + a_1^3a_i\mathcal{P}_1(\vec{\Omega}) + a_1^2a_i^2\mathcal{P}_2(\vec{\Omega}) + a_1a_i^3\mathcal{P}_3(\vec{\Omega}) + a_i^4\mathcal{P}_4(\vec{\Omega}) \\ &+a_1^2\mathcal{P}_{\text{bkgint}}^{0}(\vec{\Omega}) + a_1a_i\mathcal{P}_{\text{bkgint}}^{1}(\vec{\Omega}) + a_i^2\mathcal{P}_{\text{bkgint}}^{2}(\vec{\Omega}) \\ +& \mathcal{P}_{\text{bkg}}(\vec{\Omega})\end{aligned} \tag{5.11}$$

Some background processes, such as QCD-induced qq → ZZ, do not interfere with signal and are included separately in Eq. (5.3) as before.

The offshell events are divided into categories using the same criteria as in the previous section, and similar discriminants are used. In the onshell region, $\mathcal{D}_{\text{bkg}}$ combined the four lepton invariant mass with the other kinematic information, because signal events are expected to have an invariant mass of 125 GeV ± detector resolution. In the offshell region, the invariant mass is nowhere near 125 GeV, and the shape of the mass spectrum can provide additional information to measure the width and anomalous couplings. Therefore, the mass is used as a separate observable, and the other kinematic information to separate signal from background is separated into another observable $\mathcal{D}^{\text{kin}}_{\text{bkg}}$, which includes decay information in all categories and VBF or VH information in the respective categories. Additionally, a pure discriminant, such as $\mathcal{D}_{0-}$ in the case of the f_{a3} measurement, separates the SM from the anomalous contribution.

The discriminants used in the three categories, again using the example of the f_{a3} analysis, are shown in Fig. 5.10.

The offshell anomalous coupling results are shown in Fig. 5.11. The improvement brought by the offshell region is primarily illustrated by the difference between the green curves, which use only onshell events, and the red ones, which use offshell events as well while allowing the width to float.

Figure 5.12 shows the likelihood scan for the Higgs boson width. To make a more model-independent measurement, various configurations are used for the fit, each one floating a different anomalous coupling. No matter which coupling is floated, the results are the same: the zero width hypothesis for the Higgs boson width is excluded at 95% confidence level, as is a width 2.2 times larger than the SM width.

5.4 High Mass Search

Using similar methods to the offshell analysis, it is possible to search for a new resonance in the high mass region [28]. This resonance could have a significant width, which would mean that it interferes with background and with the offshell tail of the Higgs boson, as described in Sect. 4.5.2. Additionally, it could be produced through any combination of gluon fusion and VBF. The high mass search uses both the 4ℓ, $2\ell 2q$, and $2\ell 2\nu$ final states. In the 4ℓ channel, 3 categories are used: untagged, VBF-tagged, and RSE. The RSE category, which stands for "reduced selection electrons", includes events with electrons that fail some of the normal selection criteria, which can be bypassed in the high mass region due to lower background. The categorization schemes are different for the other final states, but all cases include a category targeting VBF events and a category targeting gluon fusion events (Fig. 5.13).

The results of the analysis are shown in Fig. 5.14. No new resonance is found.

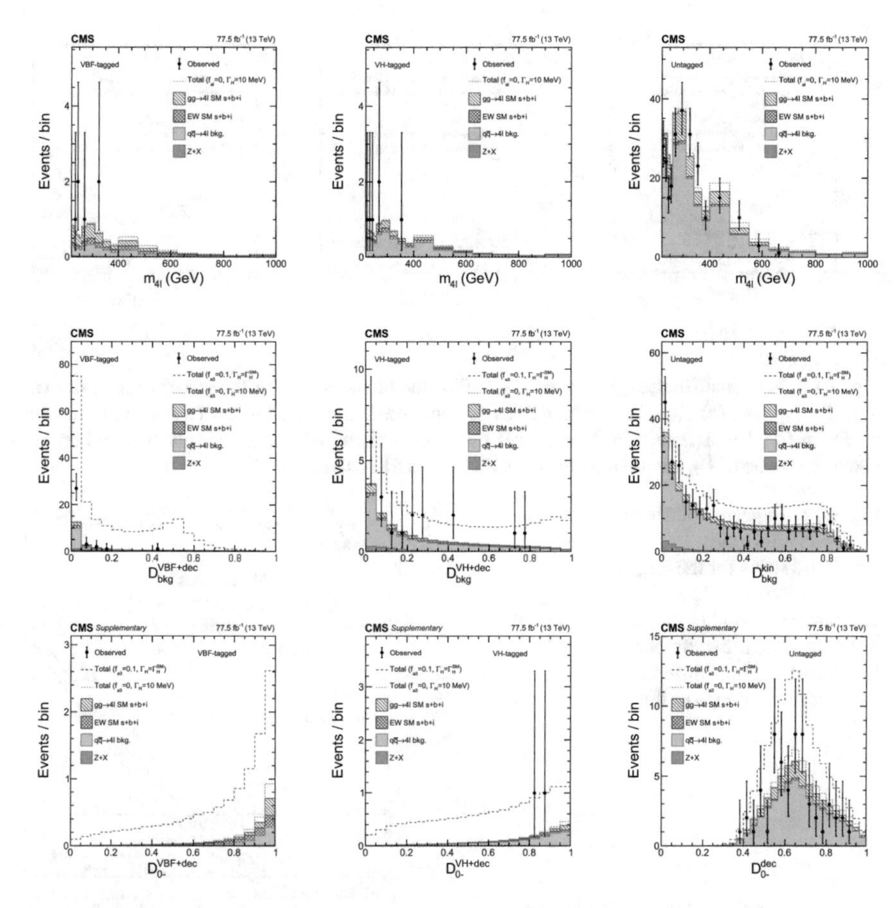

Fig. 5.10 Distributions of the three discriminants used to measure f_{a3} in the three categories of the offshell region. The top row shows $m_{4\ell}$, the middle row shows $\mathcal{D}^{\text{kin}}_{\text{bkg}}$, and the bottom row shows $\mathcal{D}_{0-}$. The type of kinematic information used for each discriminant depends on the category. The left column shows the discriminants in the VBF-tagged category, the middle column shows the ones in the VH-tagged category, and the right column shows the discriminants for the untagged category. To enhance the signal over background contributions, all of the plots except $\mathcal{D}_{\text{bkg}}$ use a cut $\mathcal{D}_{\text{bkg}} > 0.6$, and all of the plots except $m_{4\ell}$ use a cut $m_{4\ell} > 340$ GeV [27]

5.5 Anomalous Couplings in the H → ττ Channel

Searches for anomalous HVV couplings using decay information are limited to the H → ZZ and H → WW decays. However, searches for anomalous couplings in production can happen in any decay channel. This section will discuss results from CMS's anomalous couplings analysis in H → ττ, using data from 2016 [29].

Detecting the ττ final state and separating it from background requires different analysis methods than the ZZ → 4ℓ final state used in the rest of the analyses here. This analysis closely follows the methods used for the discovery of the H → ττ

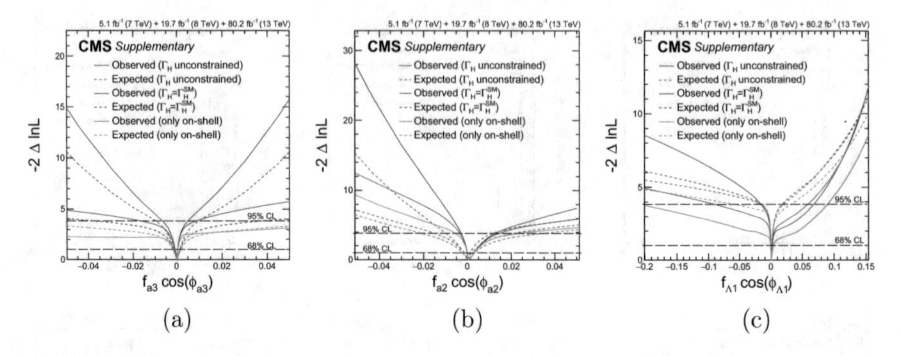

Fig. 5.11 Expected (dashed) and observed (solid) likelihood scans for the effective fractions (**a**) f_{a3}, (**b**) f_{a2}, and (**c**) $f_{\Lambda 1}$. The green curves use only onshell events and are equivalent to the red curves in Fig. 5.9. The red and blue curves use both onshell and offshell events. The red curves allow any value of Γ_H, while the blue ones fix it to its SM expectation [27]

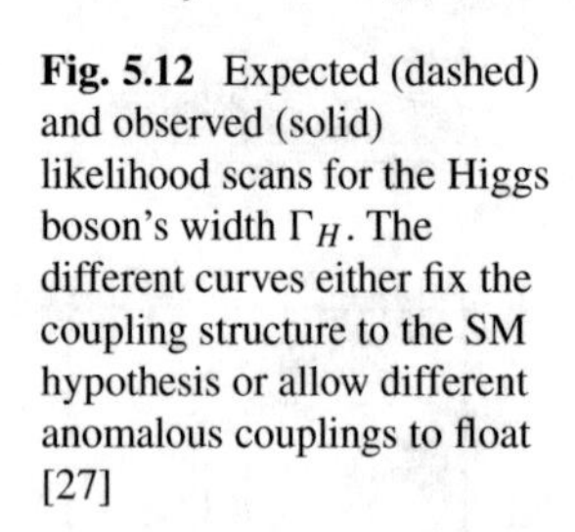

Fig. 5.12 Expected (dashed) and observed (solid) likelihood scans for the Higgs boson's width Γ_H. The different curves either fix the coupling structure to the SM hypothesis or allow different anomalous couplings to float [27]

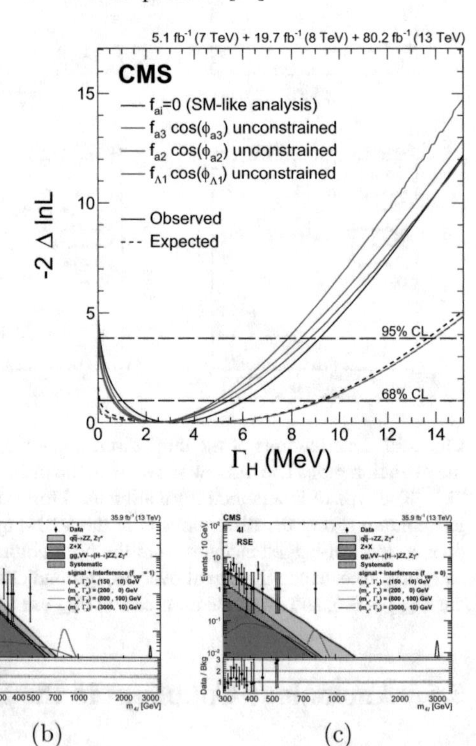

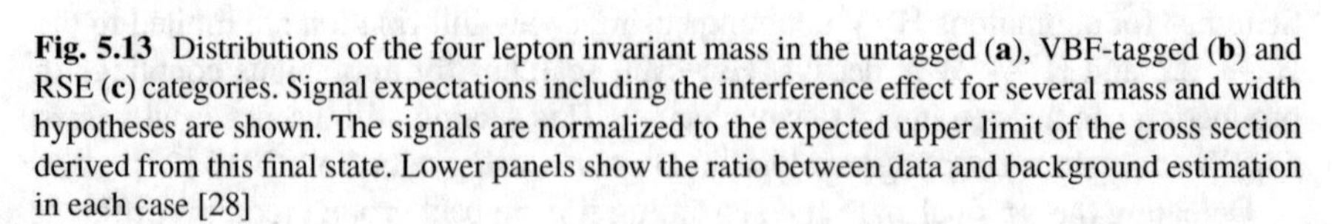

Fig. 5.13 Distributions of the four lepton invariant mass in the untagged (**a**), VBF-tagged (**b**) and RSE (**c**) categories. Signal expectations including the interference effect for several mass and width hypotheses are shown. The signals are normalized to the expected upper limit of the cross section derived from this final state. Lower panels show the ratio between data and background estimation in each case [28]

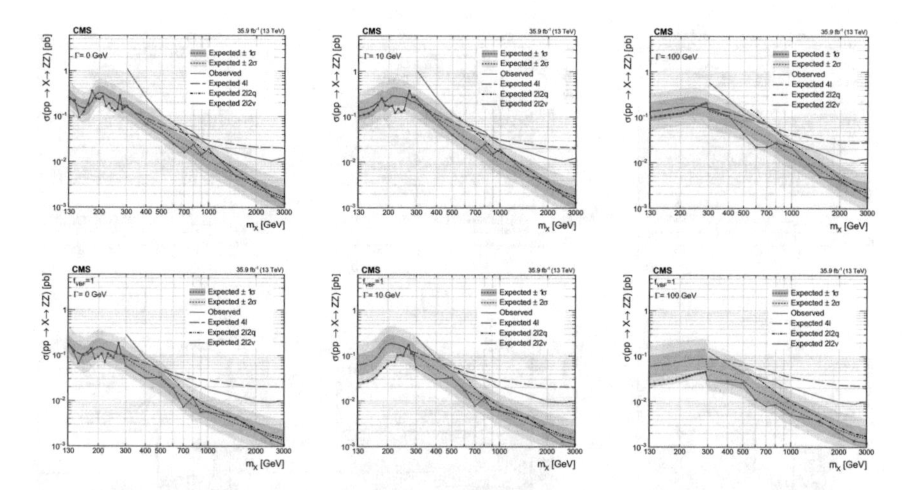

Fig. 5.14 Expected and observed upper limits at 95% CL on the pp → X → ZZ cross section as a function of m_{X} and for several Γ_{X} values with f_{VBF} as a free parameter (top row) and fixed to 1 (bottom row). The results are shown for 4ℓ, $2\ell 2$q, and $2\ell 2\nu$ channels separately and combined. The reported cross section corresponds to the signal-only contribution in the absence of interference [28]

decay by CMS [30]. The events are divided into categories based on how the τ leptons decay, in categories called $\tau_h\tau_h$, eτ_h, $\mu_h\tau_h$, and eμ. The τ_h decays include all hadronic decays, typically including various pions and kaons. All τ_h decays include a neutrino, and leptonic decays include two, so the reconstruction is complicated by the fact that neutrinos can only be reconstructed through MET. The other possible final states, ee and $\mu\mu$, are not included due to the overwhelming background in those channels. The algorithm used to identify hadronic τ decays is described in [31, 32].

Because there is no HVV vertex on the decay side, the gluon fusion process is unchanged for any anomalous couplings. The VBF and VH processes have a single HVV vertex, and their shape as a function of anomalous couplings is described by Eq. (5.2), just like a H → ZZ → 4ℓ decay produced in gluon fusion.

The events are divided into three categories:

- The 0-jet category contains events with no jets that have $p_T^{\tau\tau} > 30$ GeV.
- The VBF category contains events with two jets that satisfy various requirements to isolate the VBF topology. These cuts vary by category in order to suppress category-specific backgrounds, but typically require a large dijet invariant mass m_{JJ}, a large η separation between the jets, and/or a high $p_T^{\tau\tau}$ invariant mass.
- The boosted category includes all events that do not fall into the other two categories. It is called "boosted" because the events have at least one jet, giving the H boson a nonzero p_T.

The backgrounds for this analysis are complicated enough that there is no simple way to construct a $\mathcal{D}_{\mathrm{bkg}}$ observable. Instead, we use the mass of the visible τ decay

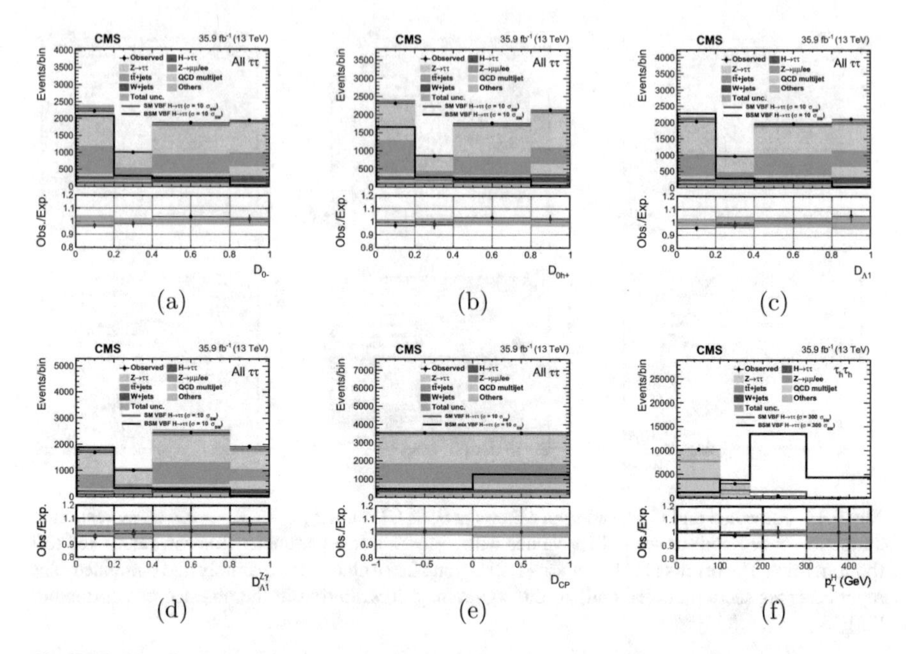

Fig. 5.15 Distributions of the discriminants used to measure anomalous couplings in the H $\to \tau\tau$ final state: (**a**) $\mathcal{D}_{0-}$ for f_{a3}, (**b**) $\mathcal{D}_{0h+}$ for f_{a2}, (**c**) $\mathcal{D}_{\Lambda1}$ for $f_{\Lambda1}$, (**d**) $\mathcal{D}_{\Lambda1}^{Z\gamma}$ for $f_{\Lambda1}^{Z\gamma}$. (**e**) shows $\mathcal{D}_{CP}$, used to detect interference between a_1 and a_3, and (**f**) shows the p_T distribution in the boosted category [29]

products m_{vis} and an estimate of the actual $\tau\tau$ mass $m_{\tau\tau}$, obtained using the SVFIT algorithm [33]. In the boosted category, $p_T^{\tau\tau}$ is used, and this observable provides extra sensitivity to anomalous couplings. In the VBF category, MELA discriminants are constructed to separate between the SM and anomalous hypotheses, using information from VBF kinematics.

Some of the distributions are shown in Fig. 5.15. Because of limited statistics in control regions in data, the number of bins is reduced with respect to the analyses described earlier. However, we use the fact that the distribution of $\mathcal{D}_{CP}$ is symmetric for any CP-conserving process, which includes all backgrounds. In this way, a 2-bin distribution of $\mathcal{D}_{CP}$ can be constructed for free, without losing any statistics: it is flat for everything except the CP-violating interference components, as shown in Fig. 5.15e.

The boosted category does not have 2 VBF-like jets, and so there is not enough information to construct a MELA discriminant. However, because anomalous couplings are enhanced at higher q^2, they also result in a harder p_T spectrum. The boosted category significantly enhances the sensitivity to anomalous couplings, even when it is missing some jet information, as shown in Fig. 5.15f.

The results of this analysis are also combined with the ones from the H $\to$ VV $\to 4\ell$ analysis, described in Sect. 5.2. In doing this combination, the κ_τ coupling is allowed to float as a free parameter, so that there are three independent

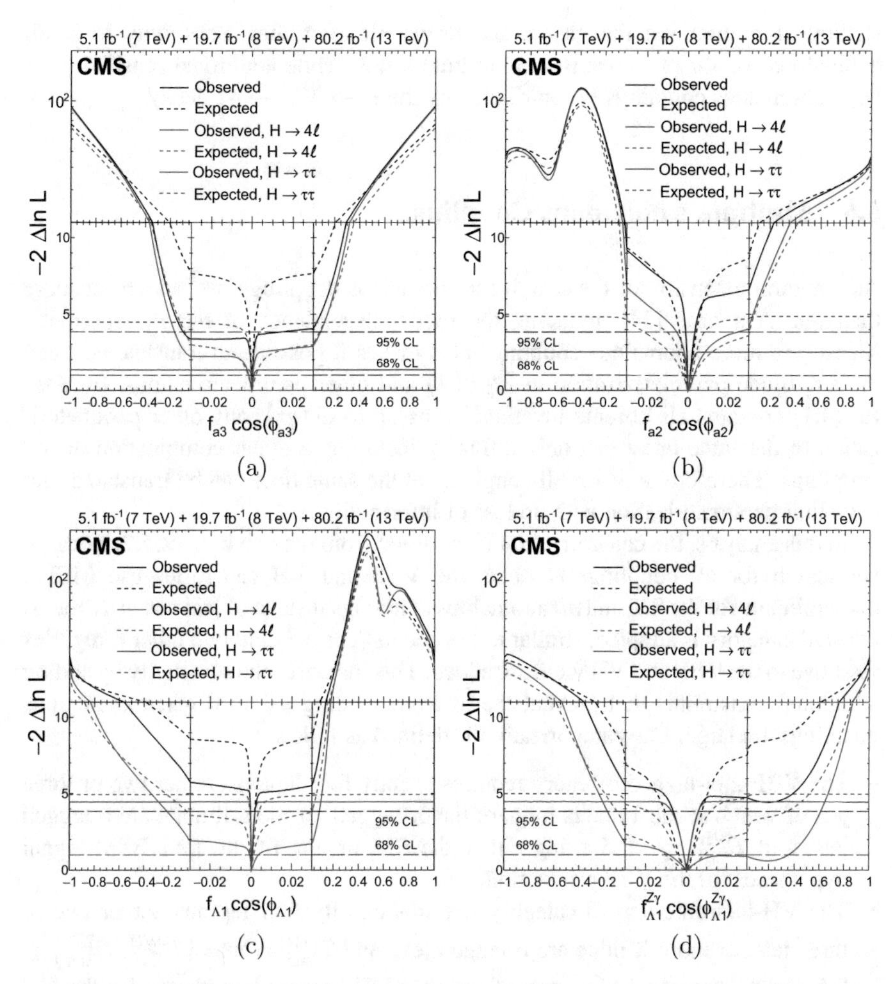

Fig. 5.16 Expected (dashed) and observed (solid) likelihood scans for the effective fractions (**a**) f_{a3}, (**b**) f_{a2}, (**c**) $f_{\Lambda1}$, and (**d**) $f_{\Lambda1}^{Z\gamma}$ from VBF and VH production and H → ZZ → 4ℓ decay information from $\tau\tau$ and four-lepton events, with all other anomalous couplings fixed to 0 [29]

parameters describing the signal strengths of different processes. The four possible μ's are related by

$$\frac{\mu_V^{\mathrm{HZZ}}}{\mu_F^{\mathrm{HZZ}}} = \frac{\mu_V^{\mathrm{H}\tau\tau}}{\mu_F^{\mathrm{H}\tau\tau}}.$$

This constraint adds additional sensitivity to the result.

The are shown in Fig. 5.16, separately and also combined with the ones from H → VV → 4ℓ. The red curves here are equivalent to the ones in Fig. 5.9. Because the $\tau\tau$ process has no decay information, it contains *only* the narrow, shallow minimum i6n the center, but levels off after that. Sensitivity to small anomalous couplings comes from both final states, while additional sensitivity for large anomalous couplings is contributed by the $\mathrm{H} \to \mathrm{VV} \to 4\ell$ decay.

5.6 Multiple Anomalous Couplings

As a natural extension of the search for anomalous couplings, we search for more than one at a time [34], reducing the model dependence of our measurement. Measuring more anomalous couplings also makes it possible to translate between the amplitude parameterization in Eq. (4.1) and other parameterizations. Because Eq. (4.1) contains all Lorentz-invariant terms up to $\mathcal{O}\left(q^2\right)$, any other parameterization to the same order can only differ by including a linear combination of our couplings. Therefore, a fit for all couplings at the same time can be translated into any other parameterization with no loss of information.

In this analysis, the categorization is modified from the one in Sect. 5.2. Because we search for all couplings at once, the VBF and VH categories use MELA discriminants for the SM and *all* anomalous couplings instead of just one at a time. A boosted category is adopted, similar to the one in Sect. 5.5, and two other categories sensitive to the VBF and VH yield are added. This increases the sensitivity by adding additional constraints that prevent the fit from sending μ_V to 0 when anomalous couplings are large. The categorization is defined as follows:

- The VBF-2jet-tagged category requires exactly four leptons, either two or three jets of which at most one is b-quark flavor-tagged, or more if none are b-tagged jets, and $\mathcal{D}_{\text{2jet}}^{\text{VBF}} > 0.5$ using either the SM or any of the four BSM signal hypotheses for the VBF production.
- The VH-hadronic-tagged category requires exactly four leptons, either two or three jets, or more if none are b-tagged jets, and $\mathcal{D}_{\text{2jet}}^{\text{VH}} = \max\left(\mathcal{D}_{\text{2jet}}^{\text{WH}}, \mathcal{D}_{\text{2jet}}^{\text{ZH}}\right) > 0.5$ using either the SM or any of the four BSM signal hypotheses for the VH production.
- The VH-leptonic-tagged category requires no more than three jets and no b-tagged jets and exactly one additional lepton or pair of opposite-sign-same-flavor leptons. In addition, events with no jets and at least one additional lepton are included in this category.
- The VBF-1jet-tagged category requires exactly 4 leptons, exactly 1 jet, and $\mathcal{D}_{\text{1jet}}^{\text{VBF}} > 0.7$. This discriminant is calculated using the SM hypothesis for the VBF production.
- The Boosted category requires exactly 4 leptons, three or fewer jets, or more if none are b-tagged jets, and the transverse momentum of the four-lepton system $p_T > 120$ GeV
- The Untagged category consists of the remaining events.

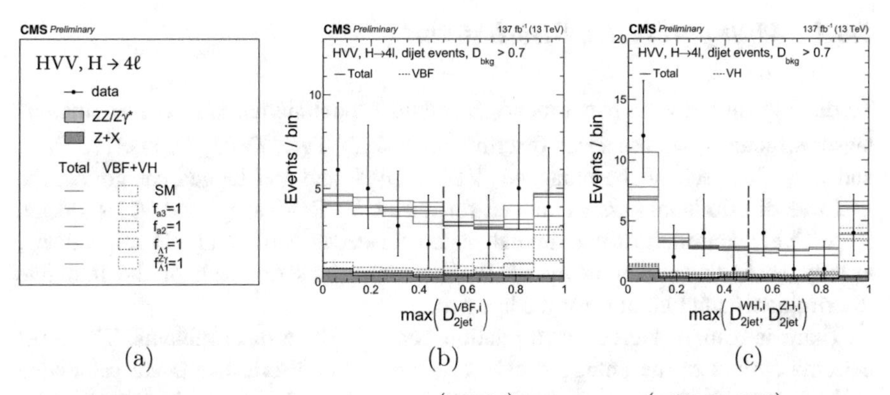

Fig. 5.17 The distributions of events for max $\left(\mathcal{D}_{\text{2jet}}^{\text{VBF},i}\right)$ (**b**) and max $\left(\mathcal{D}_{\text{2jet}}^{\text{WH},i}, \mathcal{D}_{\text{2jet}}^{\text{ZH},i}\right)$ (**c**). Only events with at least two reconstructed jets are shown, and the requirement $\mathcal{D}_{\text{bkg}} > 0.7$, where $\mathcal{D}_{\text{bkg}}$ is calculated using decay information only, is applied in order to enhance the signal contribution over the background. The VBF (**b**) and VH (**c**) signal under the SM and the four pure anomalous hypotheses, as described in the legend (**a**), is enhanced in the region above 0.5, indicated with the vertical dashed line [34]

The category discriminants, defined as the maximum of the individual discriminants for the SM and anomalous hypotheses, are shown in Fig. 5.17.

Fitting for more than one anomalous coupling at the same time essentially involves the same procedure as fitting for only one. Three additional complications arise:

1. To distinguish between several different hypotheses, more discriminants (Sect. 4.6.2) are needed.
2. As the number of couplings grows, the number of interference terms grows even faster.
3. The multidimensional fit naturally grows more complicated when there are more dimensions, especially when there are correlations between the different parameters of the fit.

In the most general case, there can be 13 parameters: a_1, a_2, a_3, and Λ_1 for ZZ and WW; a_2, a_3, and Λ_1 for Zγ; and a_2 and a_3 for $\gamma\gamma$. For this analysis, in addition to the Standard Model coupling a_1, we search for four anomalous couplings at the same time: a_2, a_3, Λ_1, and $\Lambda_1^{Z\gamma}$. As in the fits previously described, we assume that $a_i^{ZZ} = a_i^{WW} \equiv a_i$. The difficulties listed above prove to be surmountable for this fit. This procedure could be extended to search for even more couplings at a time, but the difficulties grow quickly with the number of couplings, and I will give examples for the most general case where they are relevant.

5.6.1 Multiparameter Discriminant

To distinguish between background, Standard Model signal, and four anomalous tensor structures, we use seven discriminants: $\mathcal{D}_{\text{bkg}}$, $\mathcal{D}_{0-}$, $\mathcal{D}_{0h+}$, $\mathcal{D}_{\Lambda1}$, $\mathcal{D}_{\Lambda1}^{Z\gamma}$, $\mathcal{D}_{CP}$, and $\mathcal{D}_{\text{int}}$. For each of the untagged, VBF tagged, and VH tagged categories, the first five discriminants are calculated for decay, VBF + decay, and VH + decay, respectively, and the last two are calculated for decay, VBF, and VH respectively, exactly as in the earlier analyses. We use three bins for each of the first five discriminants and two bins for the last two.

There is a high degree of correlation between these discriminants. The most extreme case is in the untagged category, where the discriminants are calculated using decay information only. As described in Sect. 4.6.2, the last six discriminants are calculated based on only five parameters: θ_1, θ_2, Φ, m_1, and m_2. ($\mathcal{D}_{\text{bkg}}$ also contains information from $m_{4\ell}$, θ^*, and Φ_1.) In the limit of an infinite number of bins, one of these discriminants is redundant. With a finite number of bins, and especially with only two or three bins as we use, each discriminant provides some information, but the correlations mean that many bins are empty. The same is true in the VBF and VH tagged categories, even though more observables go into those discriminants.

Because this analysis uses only two bins for $\mathcal{D}_{CP}$, these bins can be populated with no loss of statistics, as mentioned in Sect. 5.5. $\mathcal{D}_{\text{bkg}}$ uses more observables and is also less correlated with the other discriminants. We therefore first look at the distribution of the remaining five discriminants. This five-dimensional distribution contains $3^4 \times 2 = 162$ bins. For each category, around half of these bins contain almost no events for any signal or background process. To avoid statistical fluctuations while keeping all events as part of the measurement, all of those bins are merged into a single bin. The unrolled five-dimensional distribution contains about 80 bins taken from the original five-dimensional distribution, plus one bin that covers all of the remaining original bins.

Once the bins to be merged are identified, the final distribution to be used in the analysis contains three dimensions: $\mathcal{D}_{\text{bkg}}$ in three bins, $\mathcal{D}_{CP}$ in two bins, and the distribution described in the previous paragraph. It contains around 480 bins; however, as in Sect. 5.5, only half of that number are statistically independent due to the symmetry of $\mathcal{D}_{CP}$.

In the three new categories, boosted, VBF-1jet-tagged, and VH-leptonic-tagged, we use p_T and D_{bkg}, similar to the boosted category in Sect. 5.5.

Figures 5.18, 5.19, 5.20, and 5.21 shows distributions of the discriminants used for the analysis.

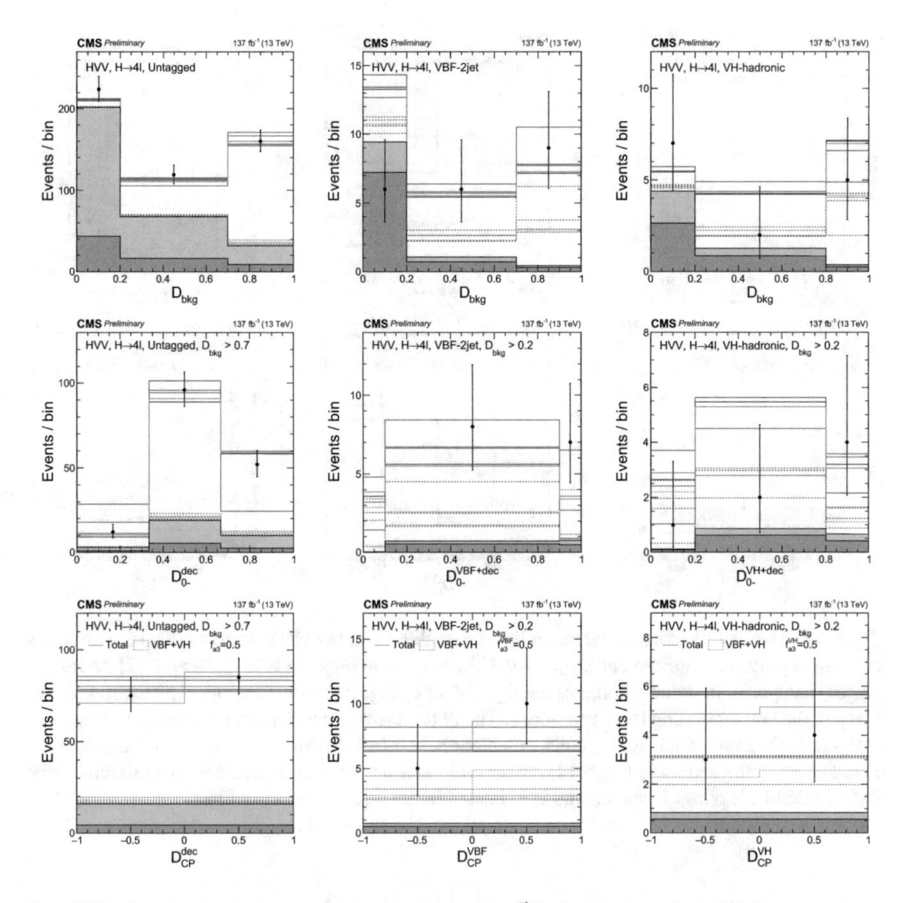

Fig. 5.18 The distributions of events in the observables $\vec{x}$ in the HVV analysis. The top row shows $\mathcal{D}_{\text{bkg}}$ in the VBF-2jet-tagged (left), VH-hadronic-tagged (middle), and untagged (right) categories. The rest of the distributions are shown with the requirement $\mathcal{D}_{\text{bkg}} > 0.7(0.2)$ in the untagged (VBF-2jet- and VH-hadronic-tagged) categories in order to enhance the signal over background contributions. The middle row shows $\mathcal{D}_{0-}$ in the corresponding three categories. The bottom row shows $\mathcal{D}_{CP}$ in the corresponding three categories. Observed data, background expectation, and five signal models are shown on the plots as indicated in the legend in Fig. 5.17a. In several cases a sixth signal model with a mixture of the SM and BSM couplings is shown and is indicated in the legend explicitly [34]

5.6.2 *Template Parameterization*

The primary difficulty of the multiparameter analysis is the number of templates, or histograms, needed to parameterize the probability distribution grows quickly with number of couplings. For processes with a single HVV vertex, such as $gg \to H \to ZZ$, the probability distribution is

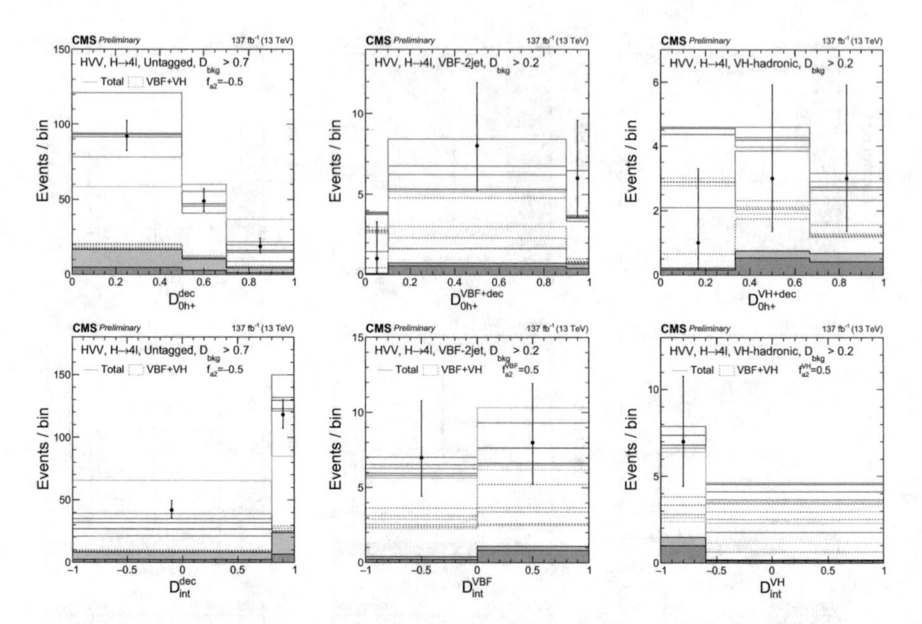

Fig. 5.19 The distributions of events in the observables $\vec{x}$ in the HVV analysis. The distributions are shown with the requirement $\mathcal{D}_{\text{bkg}} > 0.7(0.2)$ in the untagged (VBF-2jet- and VH-hadronic-tagged) categories in order to enhance the signal over background contributions. The top row shows $\mathcal{D}_{0h+}$ in the corresponding three categories. The bottom row shows $\mathcal{D}_{\text{int}}$ in the corresponding three categories. Observed data, background expectation, and five signal models are shown on the plots as indicated in the legend in Fig. 5.17a. In several cases a sixth signal model with a mixture of the SM and BSM couplings is shown and is indicated in the legend explicitly [34]

$$P(a_i, \vec{\Omega}) \sim \left| \sum_{i=1}^{N} a_i A_i \left(\vec{\Omega}\right) \right|^2, \tag{5.12}$$

where $\vec{\Omega}$ is the observables, A_i is the amplitude corresponding to the coupling a_i and N is the number of couplings in the fit, including the Standard Model coupling a_1. When multiplied out, assuming the couplings are real, the probability distribution contains $\binom{N+2-1}{2}$ terms that look like $a_i a_j T_{ij}\left(\vec{\Omega}\right)$, where $T_{ij}\left(\vec{\Omega}\right)$ is a product of amplitudes and is parameterized by a template. In the four-anomalous-coupling fit described here, this number is 15. In the most general case with 13 parameters, we have $\binom{9+2-1}{2} = 45$ terms, because the WW couplings do not contribute to the 4ℓ decay.

The number of templates grows significantly faster when we consider a process with two HVV vertices, such as VH or VBF production. In this case, the probability distribution is

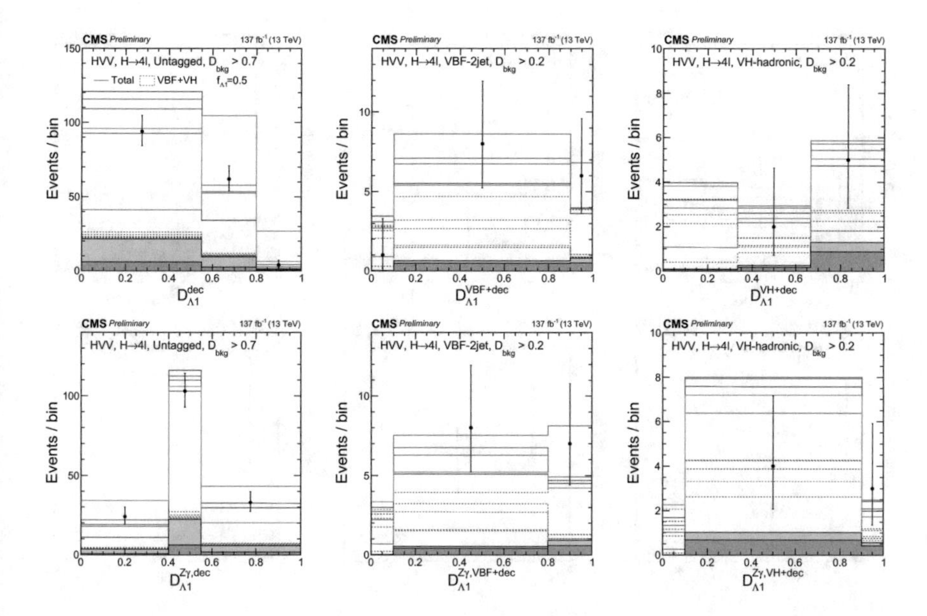

Fig. 5.20 The distributions of events in the observables $\vec{x}$ in the HVV analysis. The distributions are shown with the requirement $\mathcal{D}_{\text{bkg}} > 0.7(0.2)$ in the untagged (VBF-2jet- and VH-hadronic-tagged) categories in order to enhance the signal over background contributions. The top row shows $\mathcal{D}_{\Lambda 1}$ in the corresponding three categories. The bottom row shows $\mathcal{D}_{\Lambda 1}^{Z\gamma}$ in the corresponding three categories. Observed data, background expectation, and five signal models are shown on the plots as indicated in the legend in Fig. 5.17a. In several cases a sixth signal model with a mixture of the SM and BSM couplings is shown and is indicated in the legend explicitly [34]

$$P(a_i, \vec{\Omega}) \sim \left| \sum_{i=1}^{N} \left[a_i A_i^{\text{VBF}} \left(\vec{\Omega} \right) \right] \sum_{i=1}^{N} \left[a_i A_i^{\text{decay}} \left(\vec{\Omega} \right) \right] \right|^2, \tag{5.13}$$

which multiplies out to $\binom{5+4-1}{4} = 70$ terms in our four-anomalous-coupling fit. Each term looks like $a_i a_j a_k a_l T_{ijkl} \left(\vec{\Omega} \right)$ and is again represented by a template. The fully general fit with 13 parameters contains 1605 terms for VBF. This number comes from a sum of binomial coefficients to address the fact that VBF production includes WW couplings and 4ℓ decay does not.

The number of templates is increased further because a separate parameterization is needed for each of the categories and lepton flavor combinations. All told, several thousand templates are needed for the four-parameter fit, and an order of magnitude more would be needed for the fully general case.

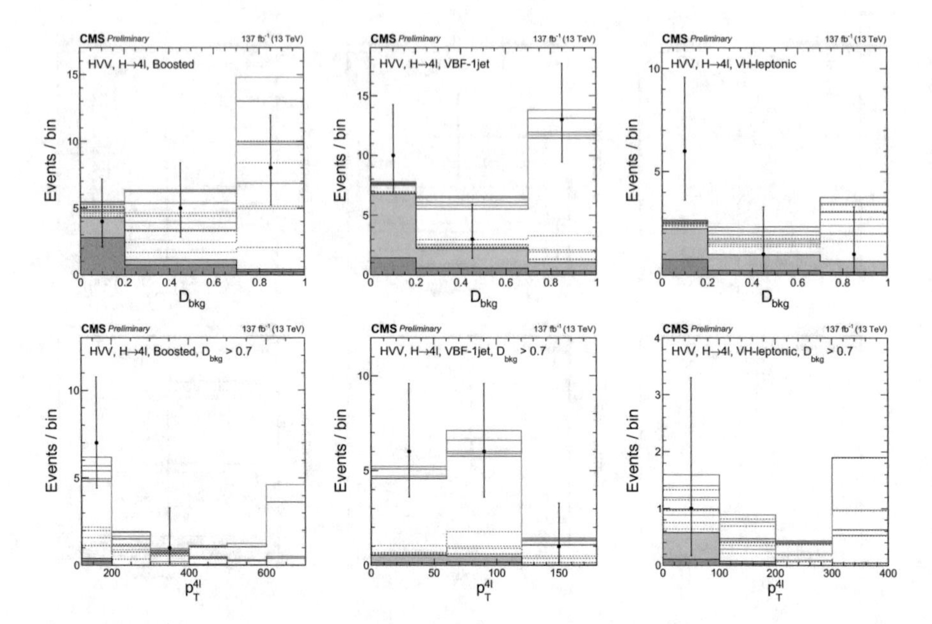

Fig. 5.21 The distributions of events in the observables $\vec{x}$ in the HVV analysis. The top row shows $\mathcal{D}_{\text{bkg}}$ in the boosted (left), VBF-1jet-tagged (middle), and VH-leptonic-tagged (right) categories. The bottom row shows $p_T^{4\ell}$ in the corresponding three categories. The $p_T^{4\ell}$ distributions are shown with the requirement $\mathcal{D}_{\text{bkg}} > 0.7$ in order to enhance the signal over background contributions. Observed data, background expectation, and five signal models are shown on the plots as indicated in the legend in Fig. 5.17a [34]

5.6.2.1 Avoiding Statistical Fluctuations

An important consideration is avoiding statistical fluctuations in the templates. Most people find it impossible to visualize a seven-dimensional distribution, and it is even more impossible to visualize thousands of seven-dimensional distributions, so visual sanity checks are not feasible. One simple check is to change the binning—for example, remove both the background contribution and $\mathcal{D}_{\text{bkg}}$. This does not appreciably change the expected limits, indicating that statistical fluctuations are small enough that they do not impact the results.

However, one type of statistical fluctuation is particularly dangerous. If an interference term fluctuates up in a particular bin and a pure term fluctuates down, it is possible that, for a particular combination of the parameters, the total probability parameterization is negative. This is physical and mathematical nonsense, and it causes the fit to fail. A safeguard is needed to avoid this behavior.

To populate the templates, we use the following algorithm, separately for each bin: for each template a, we obtain an estimate of the bin's content by reweighting, as described in Sect. 4.6.1, from each of the generated samples b: $x_{ab} \pm \delta x_{ab}$. To simplify the computation, we need to approximate the weighted Poisson distribution as a Gaussian distribution, where the error is the square root of the sum of weights squared.

Errors on a Poisson count are notoriously difficult to estimate when few statistics are available, as is the case in several of the bins. However, in our case we have a way of determining when an error estimate is too small and correcting for it: if sample b has zero or few events in a particular bin, we look at the better- or similarly-populated samples b' and inflate δx_{ab} to $\delta x_{ab'}$.

Finally, we estimate the final bin content y_a. For reasons that will be made clear below, we do this at the same time for all a by parameterizing the likelihood of a particular set of bin contents $\vec{y}$. In principle, this is a multidimensional Poisson distributions, including the *correlations* among these dimensions because the same samples are used to produce those events. However, to simplify the math and computation time, we approximate it as an uncorrelated Gaussian distribution:

$$-2\ln L(\vec{y}) = \sum_{a,b}\left(\frac{y_a - x_{ab}}{\delta x_{ab}}\right)^2 \tag{5.14}$$

In this form, maximizing the likelihood, or minimizing Eq. (5.14), is simple, as it is just a sum of quadratics. This gives us the first estimate of $\vec{y}$.

The next step is to check whether this estimate is feasible, i.e., that the probability for an event to fall in the bin remains positive for all possible values of the couplings.

Determining whether the probability parameterization can ever go negative for a particular $\vec{y}$ is a complicated undertaking and requires a section of its own.

5.6.2.2 Detecting Negative Probability

In this section, it is necessary to make the relationship between the templates explicit instead of just enumerating them. Therefore, I will expand the index a of the previous section into ij or $ijkl$ from Eqs. (5.12) and (5.13). I will first describe the simplest case, gluon fusion with a single anomalous coupling, and progress to the most complicated, VBF with multiple anomalous couplings.

For gluon fusion with a single parameter, there are only three templates, with bin contents y_{11}, y_{12}, and y_{22}. Equation (5.12) expands into

$$P(a_i, \vec{\Omega}) \sim a_1^2 y_{11} + a_1 a_2 y_{12} + a_2^2 y_{22}, \tag{5.15}$$

which is always positive as long as

$$\begin{aligned} y_{11} &> 0 \\ y_{12} &> 0 \\ \left|y_{12}\left(\vec{\Omega}\right)\right| &\leq 2\sqrt{y_{11}y_{12}} \end{aligned} \tag{5.16}$$

With multiple parameters, the criteria for the gluon fusion probability density to always be positive are similarly

$$\begin{aligned} y_{ii} &> 0 \\ \left| y_{ij}\left(\vec{\Omega}\right) \right| &\leq 2\sqrt{y_{ii} y_{jj}} \end{aligned} \tag{5.17}$$

for all $i \neq j$.

The VBF probability density is more complicated. With a single parameter, Eq. (5.13) expands into

$$P(a_i, \vec{\Omega}) \sim a_1^4 y_{1111} + a_1^3 a_2 y_{1112} + a_1^2 a_2^2 y_{1122} + a_1 a_2^3 y_{1222} + a_2^4 y_{2222}. \tag{5.18}$$

To ensure that this is always positive, we first ensure that y_{1111} and y_{2222} are positive. Then, we set $a_1 = 1$ (or, equivalently, divide through by a_1^4) to obtain a quartic polynomial $f(a_2)$. We differentiate, obtaining a cubic polynomial $f'(a_2)$, and find its 1, 2, or 3 real zeros z_i using the cubic formula. We then plug those zeros into the original quartic polynomial and find the smallest $f(z_i)$. This is the minimum of the quartic polynomial. The criteria for $\vec{y}$ to be reasonable are

$$\begin{aligned} y_{1111} &> 0 \\ y_{2222} &> 0 \\ \min_i \left(f\left(z_i\right)\right) &\geq 0 \end{aligned} \tag{5.19}$$

The most complicated case is VBF with multiple parameters. Unlike in gluon fusion, there are terms with up to four different couplings, which means that there is no way to decouple the interference terms between different couplings, as we did in the case of gluon fusion. We end up with a multivariate quartic polynomial, $\sum_{i,j,k,l} y_{ijkl} a_i a_j a_k a_l$, which is similar to Eq. (5.18) but with more terms (70, where there are four anomalous couplings). In theory, the strategy for minimizing this is the same as in the one-parameter case: set $a_1 = 1$ to obtain a quartic polynomial $f(a_2, \ldots)$, find the gradient $\vec{\nabla} f$ by differentiating with respect to each of the rest of the parameters, solve $\vec{\nabla} f = 0$, find the value of $f(a_2, \ldots)$ at each of the real solutions, and take the smallest.

Practically, the difficult part of this is solving the system of cubic equations $\vec{\nabla} f = 0$. Solving simultaneous polynomial equations in general is a complicated task.

Extensive discussion of algebraic approaches to this problem can be found in [35]. One approach is to find what is known as a Gröbner basis by means of Buchberger's algorithm [36]. In practical terms, a Gröbner basis is a set of polynomials that have the same solutions as the original ones, but with particular mathematical properties, with the result that they can be more easily solved. Gröbner bases in general are unstable with respect to small changes in the coefficients and are therefore only practical when working with integer or small rational coefficients. The algorithm in [37] produces modified Gröbner bases that are stable for floating point numbers, with control over the size of the deviation from the real Gröbner basis for the system. Unfortunately, the efficiency of running this algorithm is highly sensitive to the chosen order of terms in the polynomial; when a "bad" order is chosen, it runs for many minutes and produces several gigabytes of output before converging. In our case, we need to solve hundreds of systems of cubic polynomials. Determining the best order by trial and error is too slow, and there is no obvious structure to the coefficients that would help to determine an order. This approach is therefore not feasible for our application.

Another method for solving polynomial equations is known as homotopy continuation, described in [38]. This method is analytical rather than algebraic. It starts with a similar system of polynomials with known solutions, such as $x_i^3 - 1 = 0$ for however many i's are needed. It then continuously transforms the system, tracking the solutions in the complex space, until it reaches the one we want to solve. We use the Hom4PS program [39, 40] for this. The efficiency of running homotopy continuation is *simplified* by the fact that because our polynomials have random coefficients, they are unlikely to have degenerate roots or solutions where one of the variables is zero, cases which require special treatment. Hom4PS takes about half a second to solve the system of four cubic polynomials needed for the four-parameter fit.

At Infinity

The multidimensional case has a further complication: the polynomial can be negative at infinity. For the single parameter case in Eq. (5.22), bad behavior at infinity can be avoided by just requiring that the pure terms y_{1111} and y_{2222} are positive. For the multidimensional case, instead of two points, we have a sphere at infinity and have to avoid negative behavior anywhere around this sphere.

Written explicitly in an example with two anomalous couplings, the polynomial under consideration looks like

$$\begin{aligned} f(a_2, a_3) =& y_{1111} + y_{1112}a_2 + y_{1113}a_3 + y_{1122}a_2^2 + y_{1123}a_2a_3 \\ &+ y_{1133}a_3^2 + y_{1222}a_2^3 + y_{1223}a_2^2a_3 + y_{1233}a_2a_3^2 + y_{1333}a_3^3 \\ &+ y_{2222}a_2^4 + y_{2223}a_2^3a_3 + y_{2233}a_2^2a_3^2 + y_{2333}a_2a_3^3 + y_{3333}a_3^4 \end{aligned} \tag{5.20}$$

On the sphere at infinity, the terms with degree 3 dominate, giving

$$f(a_2, a_3) \approx y_{2222}a_2^4 + y_{2223}a_2^3a_3 + y_{2233}a_2^2a_3^2 + y_{2333}a_2a_3^3 + y_{3333}a_3^4 \tag{5.21}$$

We can then apply the same strategy as before: let $a_2 = 1$ and minimize this polynomial. If it is ever negative, at $a_2 = 1, a_3 = \alpha$, then the original polynomial is also negative at $a_2 = c, a_3 = c\alpha$ when c is large enough. We also have to look at the infinite points of this smaller polynomial. In this case, that just means ensuring that $y_{3333} > 0$; with more couplings, it is necessary to recursively find and minimize a boundary polynomial.

5.6.2.3 Avoiding Negative Probability

Now that we have a procedure to detect when negative probability can occur, we can construct the templates in a way that avoids it. This is accomplished by minimizing Eq. (5.14) subject to the constraint that the probability is always positive. This constraint involves all elements of $\vec{y}$, and so it is necessary to do a multidimensional minimization for all a at the same time.

To do this minimization, we use the cutting planes method for convex minimization [41] . This method relies on the fact that both the equation to be minimized and the region over which it is minimized are convex. Equation (5.14) is convex simply because it is a sum of independent, one-dimensional quadratics, each of which is convex in its own dimension. The fact that the constraint region is convex is less obvious. Written explicitly for the four-parameter VBF fit, the set of allowed $\vec{y}$ is:

$$\left\{ \vec{y} \in \mathbb{R}^{70} \middle| \forall \vec{a} \in \mathbb{R}^4 : \sum_{i,j,k,l} y_{ijkl} a_i a_j a_k a_l > 0 \right\} \tag{5.22}$$

A set is convex if, given two points $\vec{y}_1$ and $\vec{y}_2$ inside it, any point on the line between them also lies inside the set. Equation (5.22) can be viewed as an infinite set of individual constraints, each of which, despite being a complicated function of $\vec{a}$, is linear in $\vec{y}$. Each of these linear constraints is convex, and therefore, so is their intersection. Intuitively, because the quartic polynomials defined by $\vec{y}_{1,2}$ are always positive, as the polynomial moves linearly from one to the other, it remains always positive.

The cutting planes method works in iterations. First, we minimize Eq. (5.14) unconstrained, which is easy to do because it is a sum of independent quadratics, obtaining a solution $\vec{y}_1$. If this solution satisfies the constraint, there is nothing more to do. If not, we find a particular set of couplings $\vec{a}$ where the polynomial is negative, and choose the *linear* constraint defined by those couplings from Eq. (5.22). Then we minimize Eq. (5.14) again, using this linear constraint. The process continues, with more and more linear constraints, until eventually the minimization converges to a point that satisfies Eq. (5.22).

This procedure works because minimizing a sum of quadratics subject to a set of linear constraints is significantly easier than minimizing it subject to an arbitrary constraint. We use the CVXPY package [42, 43] interfaced to Gurobi [44] in each iteration.

Numerical Stability

Several mathematical tricks are used to make the minimization work better.

Scaling the Couplings

For effective numerical minimization, it is important that the relevant scales not diverge over too many orders of magnitude; instead, all numbers involved should be as close as possible to 1. A simple approach would be to scale each term of Eq. (5.14) by a factor k_a^2, which would be set so that the coefficient of each quadratic term is 1 and then divide the resulting y_a by k_a to get the final bin content. This would work perfectly well in the first iteration of the minimization. However, because these k_a do not relate to the structure of the polynomial in Eq. (5.22), we would also have to use y_a/k_a in finding the minimum of this polynomial, and the linear constraints defined by this minimum would still have large numbers. This procedure only moves the large numbers from one place to another in the fit without solving the underlying problem.

Instead, we compute $k_a = k_{ijkl}$ in a correlated way across all coefficients in a way that leaves the constraint unchanged. We accomplish this by noting that Eq. (5.22) can be rewritten, for any positive $\vec{\kappa}$, as

$$\left\{ \vec{y} \in \mathbb{R}^{70} \middle| \forall \vec{a} \in \mathbb{R}^4 : \sum_{i,j,k,l} (y_{ijkl}\kappa_i\kappa_j\kappa_k\kappa_l) \frac{a_i}{\kappa_i}\frac{a_j}{\kappa_j}\frac{a_k}{\kappa_k}\frac{a_l}{\kappa_l} > 0 \right\} \tag{5.23}$$

$$= \left\{ \vec{y} \in \mathbb{R}^{70} \middle| \forall \vec{a} \in \mathbb{R}^4 : \sum_{i,j,k,l} (y_{ijkl}\kappa_i\kappa_j\kappa_k\kappa_l) a_i a_j a_k a_l > 0 \right\}. \tag{5.24}$$

In other words, we can pick any five κ's, one for each coupling, and set $k_{ijkl} = \kappa_i\kappa_j\kappa_k\kappa_l$. We find the optimal κ's to use by minimizing $\sum_{ijkl} \log^2(\kappa_i\kappa_j\kappa_k\kappa_l y_{ijkl})$ for the known coefficients y_{ijkl}. By this procedure, the final coefficients that go into the fit are typically in the 10^{-2}–10^2 range, which Gurobi is able to handle without a problem.

Finding "Divergent" Minima

When Hom4PS tracks the solutions of a multidimensional polynomial, sometimes one of the solutions moves away to infinity. In that case, Hom4PS calls the solution "divergent." Divergent solutions usually result from some kind of special structure in the coefficients. In our case, the coefficients are random, and so there is almost never any such structure.

However, one possibility occasionally does give rise to a divergent solution. In later iterations of the cutting plane procedure, the successive linear constraints try to eliminate negative probability by pushing the coefficients in a particular direction.

Sometimes, the result is that a negative minimum of the original polynomial gets pushed away towards infinity. If it actually reaches infinity, then it will be detected when minimizing the boundary polynomial, as described above. However, if it reaches large but finite values of the variables, Hom4PS may give up and report the solution as divergent anyway.

Empirically, this happens occasionally and is much more likely than unlucky values of the coefficients giving a divergent complex solution or a divergent real solution that corresponds to a *maximum* of the probability. Therefore, when Hom4PS reports a divergent solution, we have to take this warning seriously and identify it.

When a false divergent solution is present, it can be made finite by "inverting" the system of polynomials. First, we introduce a variable α and homogenize the cubic polynomials by multiplying each term by a power of α so that it has degree 3. The system is now underconstrained: it has one more variable than it does equations. We then choose numbers $\vec{\beta}$ and add a linear equation

$$\beta_0 + \sum_i \beta_i a_i + \beta_\alpha \alpha = 0$$

to the system and solve with Hom4PS. Each solution $(\alpha, \vec{a})$ of the new system of polynomials corresponds to a solution $\vec{a}/\alpha$ of the original system.

A false divergent solution corresponds to large values of $\vec{a}/\alpha$: large enough that Hom4PS gave up on tracking these values. The trick here is to choose $\vec{\beta}$ in such a way that $\vec{a}$ is not that large, meaning that α will be very small. In general, choosing the correct $\vec{\beta}$ is difficult. In our case, we have a handle on the correct values. As already noted, the divergent solution most likely (1) is real and (2) corresponds to a minimum. That being the case, if it had made it all the way to infinity, it would correspond to a minimum of the boundary polynomial. Even though the false minimum apparently did *not* go all the way to infinity, we expect it to be *near* a minimum of the boundary polynomial. Therefore, we can look at those minima and choose $\vec{\beta}$ such that $\sum_i \beta_i a_i = 0$ at one of them. (β_0 is arbitrary because it corresponds to a common rescaling of $\vec{a}$ and α; β_α is arbitrary because it corresponds to a rescaling of α.)

We can note here that α is essentially taking the place of a_1, which we set to 1 earlier in the process. An actual divergent minimum would indicate that our probability goes negative when $a_1 = 0$ for some values of the other couplings. A false divergent minimum means that the probability goes negative at some point with small a_1. By introducing α, we *re*introduce a_1. The linear equation gives us an estimate of the values of the couplings at the target minimum.

Permuting the Order of Variables

As mentioned earlier, the first step in finding whether the original homogeneous polynomial (Eq. (5.18) in the one-dimensional case) goes negative was to set $a_1 = 1$. This was an arbitrary choice; we could instead have set $a_2 = 1$. This choice of

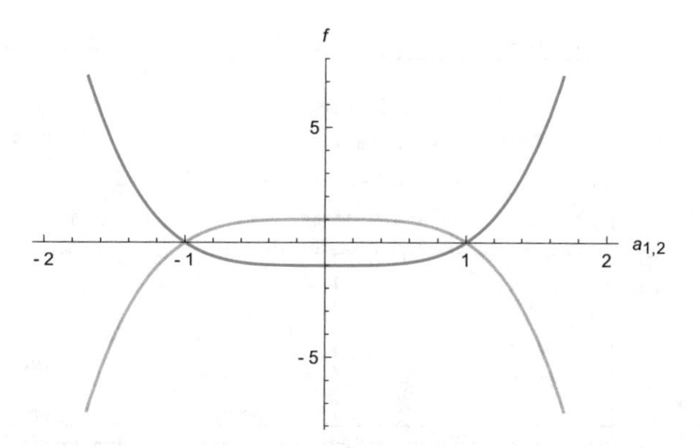

Fig. 5.22 Plots of $f(a_1) = a_1^4 - 1$ (blue) and $f(a_2) = 1 - a_2^4$ (orange). These polynomials are both obtained from the same homogeneous polynomial $f(a_1, a_2) = a_1^4 - a_2^4$ by setting one of the couplings to 1, but the places where the resulting polynomials have negative values are different

which variable to remove does not affect whether the resulting polynomial ever goes negative, but it does affect the numerical value of the constraints. For a simple example, we can take the polynomial $f(\vec{a}) = a_1^4 - a_2^4$. If we instead set $a_2 = 1$, this has a minimum of $f(0) = -1$. If we instead set $a_1 = 1$, it has no minimum, but $\lim_{a_2 \to \pm\infty} = -\infty$. These are both illustrated in Fig. 5.22.

In the one parameter case, this difference is negligible. When there are multiple parameters, some choices of which variable to remove result in large numbers in the cutting plane constraint, causing the fit to fail. The multiparameter case is also more complicated because, when looking at the behavior around the sphere at infinity, we set more variables to 1. The order in which we apply this procedure affects the numerical stability of the fit.

As long as the default procedure performs well, we ignore the potential numerical problems. If Hom4PS finds "failed paths", meaning that it loses track of one of the minima during the transformation process, we try different orders of variables until one succeeds. Similarly, when Gurobi fails for numerical reasons, we restart the cutting plane procedure. This time, when we search for negative probability in each iteration of the cutting plane method, we try every possible order of variables to remove, skipping variable orders that give divergent or failed results in Hom4PS. We choose the linear constraint that has the smallest spread in numerical values of the coefficients. In addition, we optimally look for linear constraints that involve as many as possible of the coefficients, because in our application constraints involving only a few coefficients tend to cause the cutting plane algorithm to converge slowly. This procedure is only used when necessary because each iteration is much slower than in the default procedure; however, in practice it is only needed for a few bins, so it does not significantly slow down the overall process.

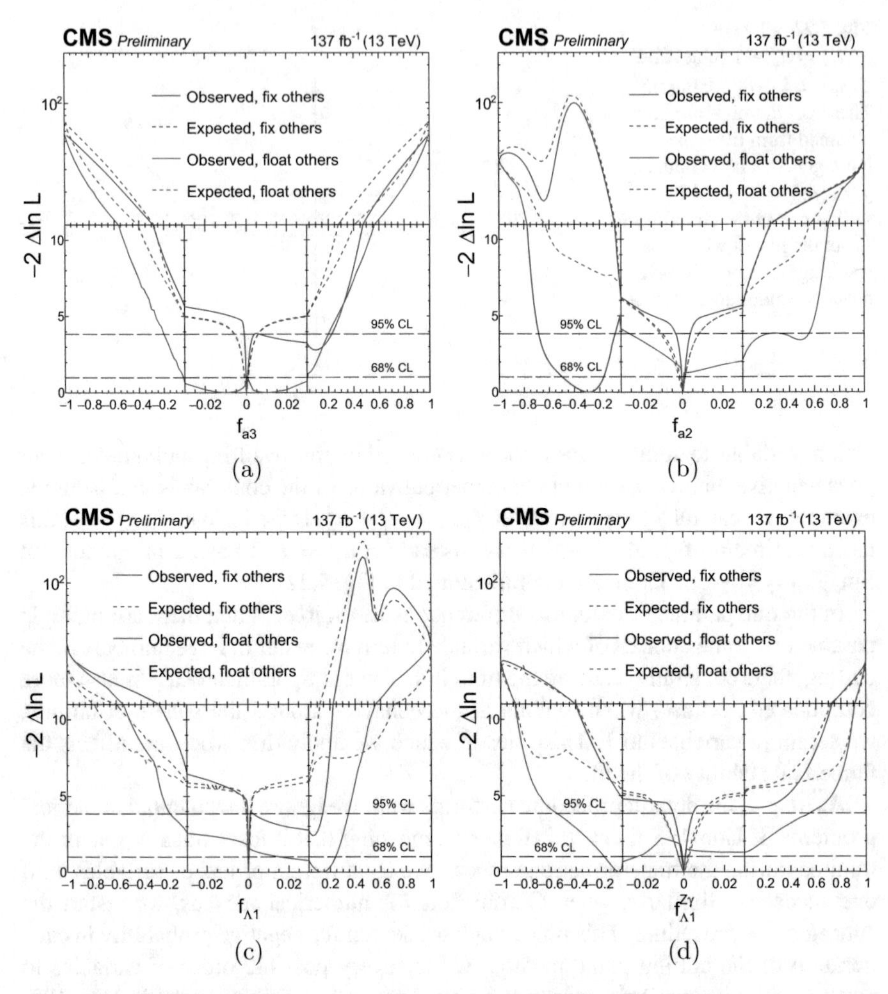

Fig. 5.23 Expected (dashed) and observed (solid) likelihood scans for the effective fractions (**a**) f_{a3}, (**b**) f_{a2}, (**c**) $f_{\Lambda 1}$, and (**d**) $f_{\Lambda 1}^{Z\gamma}$ from VBF and VH production and $\mathrm{H} \rightarrow \mathrm{ZZ} \rightarrow 4\ell$ decay information from four-lepton events, with all other anomalous couplings either fixed to 0 or floated [34]

5.6.3 Results

The results of the multiparameter HVV anomalous couplings analysis are shown in Fig. 5.23.

The blue scans in Fig. 5.23 are similar to the ones in Fig. 5.9. There are two major differences:

- The amount of data collected is increased. The previous scans used data from 2016 and 2017, and these add the data collected in 2018.

- This analysis uses additional categories, improving the sensitivity to anomalous couplings in VBF and VH production in particular.

Due to these improvements, we expect to exclude small anomalous couplings at 95% CL.

The red scans float the other anomalous couplings. The differences between red and blue are primarily at large values of the anomalous couplings, indicating that while the anomalous coupling parameters are correlated—in some cases *highly* correlated—in decay, VBF and VH production do not show these correlations, and the exclusion is about the same whether or not we float the other anomalous couplings. This is due to the fact that, as described in Sect. 4.4 and shown in Fig. 4.10, the SM is an extreme point in the parameter space with many events at low q_V^2. The anomalous couplings all show an enhancement at high q_V^2, and no correlation or interference effect will remove this enhancement. If the true minimum was at nonzero $\vec{f}_{ai}$, we would expect to see a larger difference between the blue and red curves, because $f_{a2}^{\mathrm{VBF}} = 0.1$ and $f_{a3}^{\mathrm{VBF}} = 0.1$, for example, look similar in their q^2 spectrum. Discrimination would still be available from the angles, but would be less sensitive.

The two curves meet at $f_{ai} = \pm 1$, by definition: if $f_{ai} = 1$, then all other f_{aj} must be 0.

When we look at the observed results from data, the blue curves look similar to the expectation. However, when all four anomalous couplings are allowed to float independently, the best fit values are $(f_{a3}, f_{a2}, f_{\Lambda 1}, f_{\Lambda 1}^{Z\gamma}) = (\pm 0.01, -0.29, 0.13, -0.06)$, where the two minima at positive and negative values of f_{a3} are degenerate. These global minima are driven by the decay information from $\mathrm{H} \rightarrow \mathrm{ZZ} \rightarrow 4\ell$ and is only slightly preferred to the local minimum at $(0, 0, 0, 0)$, with a difference in $-2\ln(\mathcal{L})$ of 0.1 between the SM value and the global minima. The local minimum at $(0, 0, 0, 0)$ is still evident in the four-dimensional distribution and its projections on each parameter, and is driven by the production information, as discussed above for the fits with one parameter. Due to this statistical fluctuation in the observed data when the $-2\ln(\mathcal{L})$ minima obtained from the decay and from the production kinematics differ, the observed constraints appear weaker than expected. The results are still statistically consistent with the SM and with the expected constraints in the SM.

Figure 5.24 shows the two-dimensional likelihood scans from this analysis.

5.6.3.1 EFT Relations with $SU(2) \times U(1)$ Symmetry

These studies repeated using the SU(2)×U(1) symmetry in Eqs. (4.9) to (4.13). In this case, the $f_{\Lambda 1}^{Z\gamma}$ parameter is not independent and can be derived following Eq. (4.13). Therefore, constraints on the three parameters f_{a3}, f_{a2}, $f_{\Lambda 1}$, and the signal strength are obtained in this scenario following the same approach as above. These constraints are shown in Fig. 5.25.

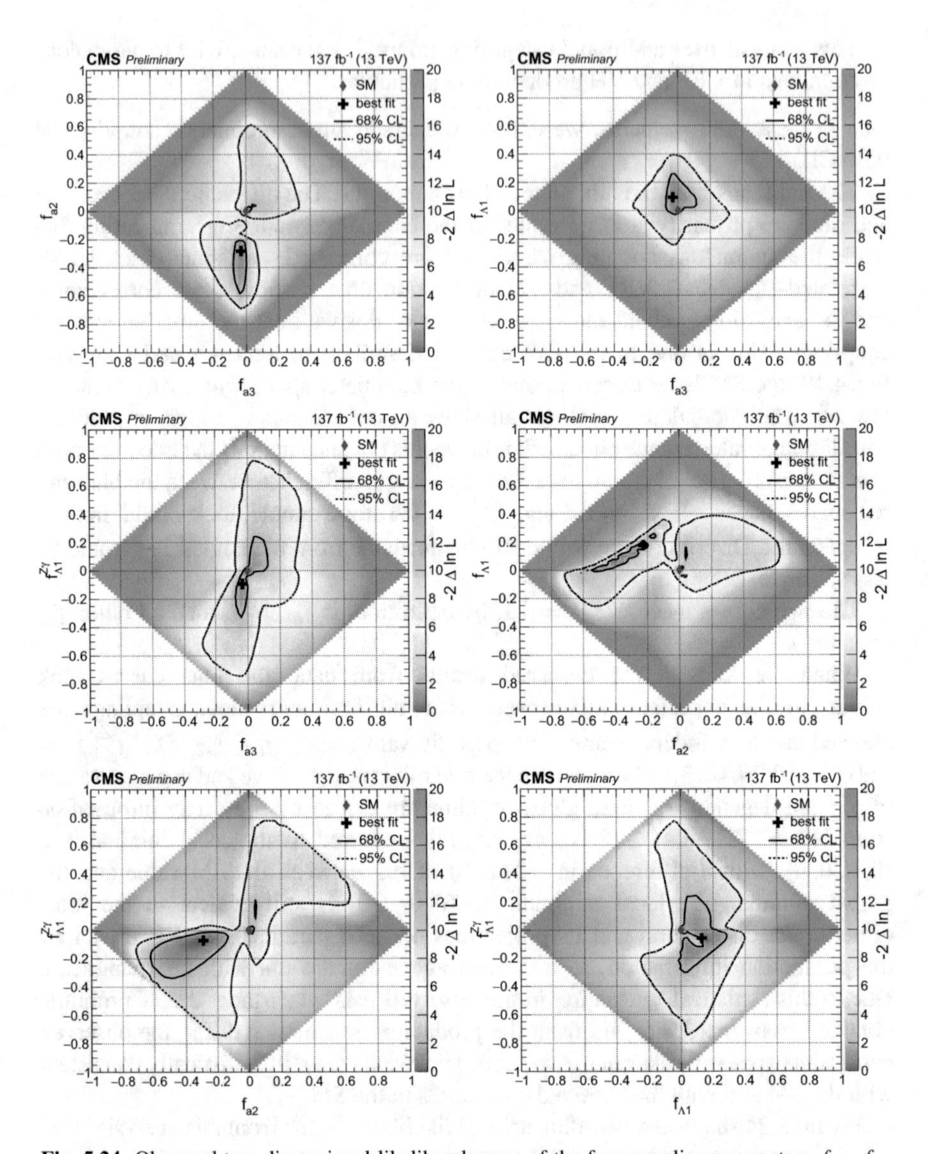

Fig. 5.24 Observed two-dimensional likelihood scans of the four coupling parameters f_{a3}, f_{a2}, $f_{\Lambda 1}$, and $f_{\Lambda 1}^{Z\gamma}$ In each case, the other two anomalous couplings along with the signal strength parameters have been left unconstrained. The 68% and 95% CL regions are presented as contours with dashed and solid black lines, respectively. The best fit values and the SM expectations are indicated by markers

Since the relationship of the HWW and HZZ couplings does not affect the measurement of the f_{a3} parameter in the H $\rightarrow 4\ell$ decay, the constraints from the decay information in the wider range of f_{a3} in Approach 2 are unaffected compared to Approach 1, when other couplings are fixed to zero. However, with

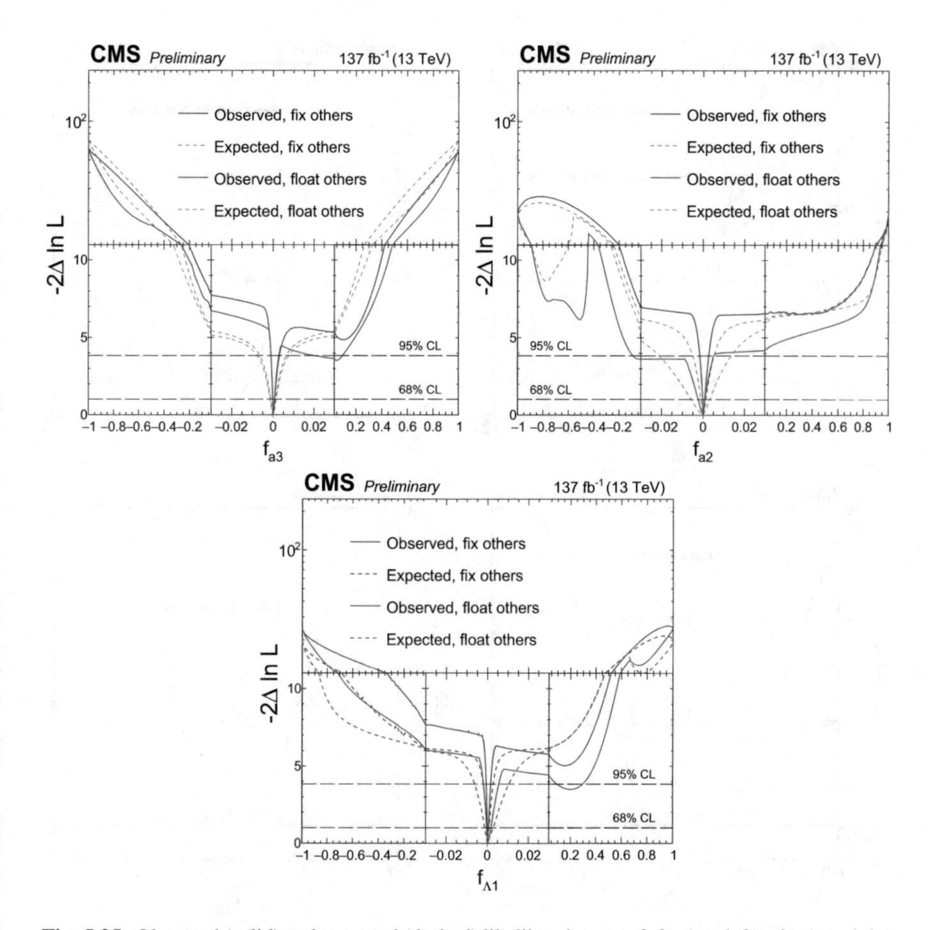

Fig. 5.25 Observed (solid) and expected (dashed) likelihood scans of f_{a3} (top left), f_{a2} (top right), and $f_{\Lambda 1}$ (bottom) with the EFT relationship of couplings set in Eqs. (4.9) to (4.13). The results are shown for each coupling separately with the other anomalous coupling fractions either set to zero or left unconstrained in the fit. In all cases, the signal strength parameters have been left unconstrained. The dashed horizontal lines show the 68 and 95% CL regions

one less parameter to float, the constraints are modified somewhat when all other couplings are left unconstrained. The modified relationship between the HWW and HZZ couplings also leads to some modification of constraints using production information in the narrow range of f_{a3}. On the other hand, the f_{a2} and $f_{\Lambda 1}$ parameters are modified substantially because the $f_{\Lambda 1}^{Z\gamma}$ information gets absorbed into these measurements.

The measurement of the signal strength μ_V and the f_{a3}, f_{a2}, and $f_{\Lambda 1}$ parameters can be re-interpreted in terms of the δc_z, c_{zz}, $c_{z\Box}$, and $\tilde{c}_{zz}$ coupling strength parameters. Observed one- and two-dimensional constraints from a simultaneous fit of EFT parameters are shown in Figs. 5.26 and 5.27. The c_{gg} and $\tilde{c}_{gg}$ couplings are left unconstrained.

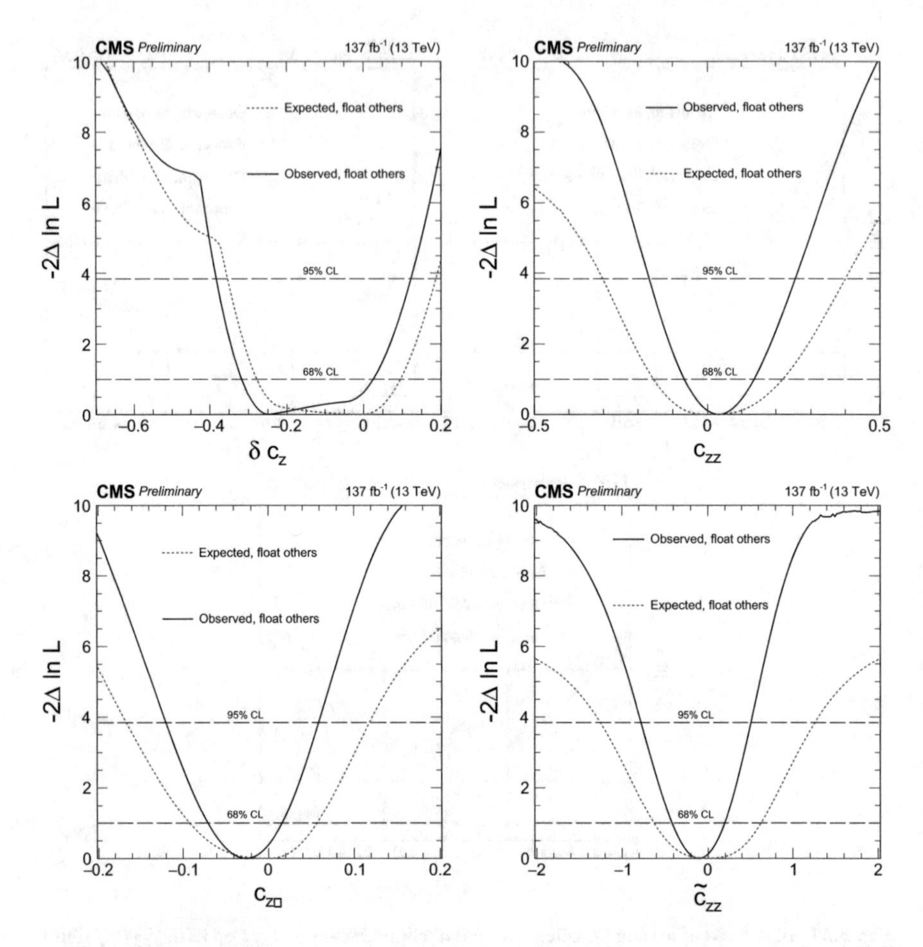

Fig. 5.26 Observed (solid) and expected (dashed) constraints from a simultaneous fit of EFT parameters δc_z (top-left), c_{zz} (top-right), $c_{z\Box}$ (bottom-left), and $\tilde{c}_{zz}$ (bottom-right) with the c_{gg} and $\tilde{c}_{gg}$ couplings left unconstrained

5.7 Hff Anomalous Couplings

This section will describe the first search for anomalous couplings in Hff. As described in Sect. 4.4.2, there is only one anomalous Hff coupling, $\tilde{\kappa}$, and it can be measured either through gluon fusion with 2 associated jets or through $t\bar{t}H$ production. This analysis uses both.

Events are divided into seven categories, similar to the ones in Sect. 5.6:

- The VBF-2jet-tagged category requires exactly four leptons, either two or three jets of which at most one is b-quark flavor-tagged, or more if none are b-tagged jets, and $\mathcal{D}_{\text{2jet}}^{\text{VBF}} > 0.5$ using the SM hypothesis.

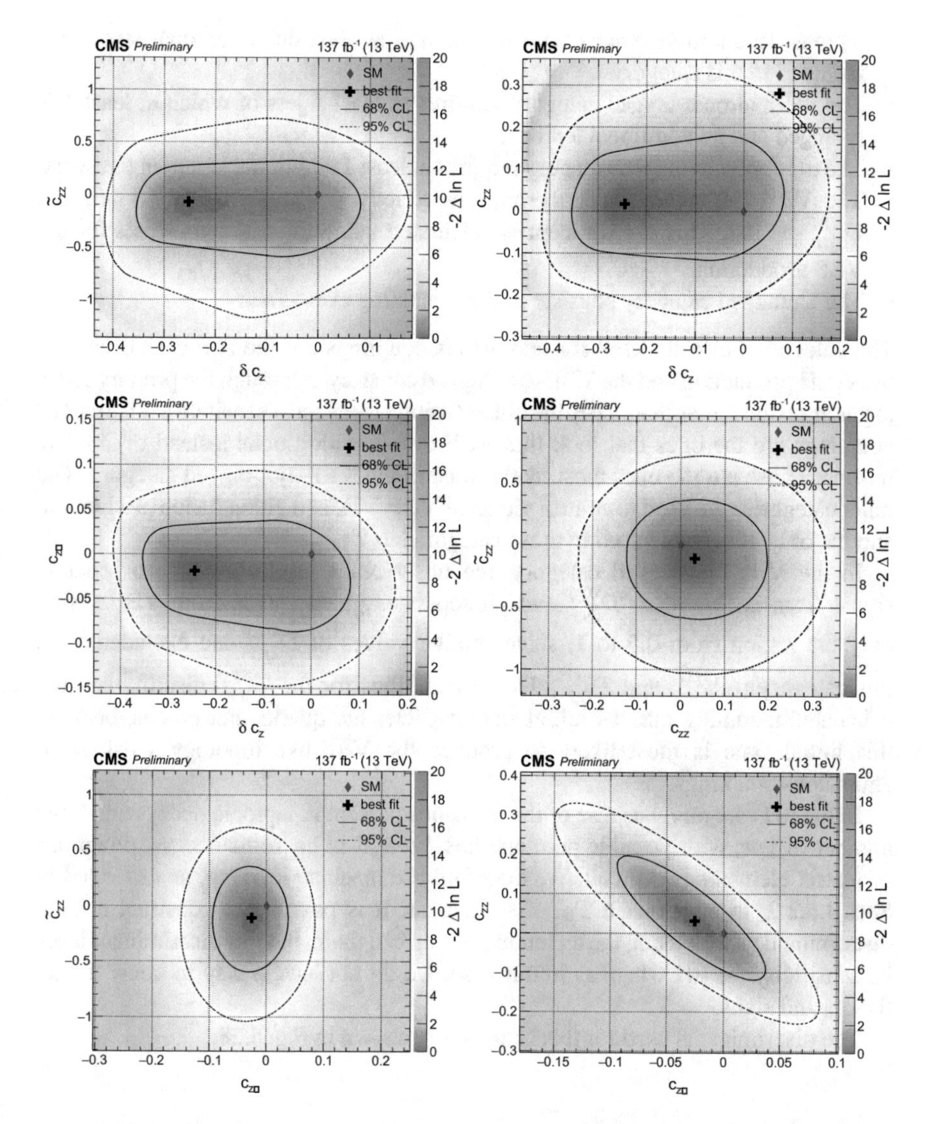

Fig. 5.27 Observed two-dimensional constraints from a simultaneous fit of EFT parameters δc_z, c_{zz}, $c_{z\Box}$, and $\tilde{c}_{zz}$ with the c_{gg} and $\tilde{c}_{gg}$ couplings left unconstrained

- The VH-hadronic-tagged category requires exactly four leptons, either two or three jets, or more if none are b-tagged jets, and $\mathcal{D}_{\text{2jet}}^{\text{VH}} = \max\left(\mathcal{D}_{\text{2jet}}^{\text{WH}}, \mathcal{D}_{\text{2jet}}^{\text{ZH}}\right) > 0.5$ using the SM hypothesis for the VH production.
- The VH-leptonic-tagged category requires no more than three jets and no b-tagged jets and exactly one additional lepton or pair of opposite-sign-same-flavor

leptons. In addition, events with no jets and at least one additional lepton are included in this category.

- The $t\bar{t}$H-hadronic-tagged category requires at least 4 jets of which at least 1 is b-tagged and no additional leptons;
- The $t\bar{t}$H-leptonic-tagged category requires at least 1 additional lepton in the event;
- The VBF-1jet-tagged category requires exactly 4 leptons, exactly 1 jet, and $\mathcal{D}_{\text{1jet}}^{\text{VBF}} > 0.7$. This discriminant is calculated using the SM hypothesis for the VBF production.
- The Untagged category consists of the remaining events.

The categories directly used for anomalous couplings are the two categories that target $t\bar{t}$H production and the VBF-2jet-tagged category. Although the primary focus is gluon fusion rather than VBF, the gluon fusion events most sensitive to anomalous couplings are the ones that look like VBF events with gluons instead of Z or W bosons. Those are the ones most likely to be in the VBF-2jet-tagged category. The other categories are used to control yields of VBF, VH, and gluon fusion production, and the only observable used in those categories is $\mathcal{D}_{\text{bkg}}$.

In the VBF-2jet-tagged category, the observables used are $\mathcal{D}_{\text{bkg}}$ to separate signal from background; $\mathcal{D}_{\text{2jet}}^{\text{VBF}}$, which separates gluon fusion from VBF (using only the region from 0.5 to 1, since smaller values of $\mathcal{D}_{\text{2jet}}^{\text{VBF}}$ are excluded from this category); $\mathcal{D}_{0-}^{\text{ggH}}$; and $\mathcal{D}_{CP}^{\text{ggH}}$. The probabilities for the $\mathcal{D}_{0-}^{\text{ggH}}$ discriminant are calculated assuming that the initial state particles are quarks, not gluons, because this initial state is most likely to produce the VBF-like topology sensitive to anomalous couplings.

In the $t\bar{t}$H category, because of the neutrinos present in leptonic decays and large number of jets, with possible permutations, present in hadronic decays, direct use of matrix elements is difficult. We therefore use machine learning, as described in Sect. 4.6.2.2, to construct a $\mathcal{D}_{0-}$ discriminant. It is possible to construct a $\mathcal{D}_{CP}$ discriminant as well using the techniques described there, but this discriminant loses its sensitivity for $t\bar{t}$H when, as in our case, we do not have a way to know the jet flavors and signs.

The discriminants used for this analysis are shown in Fig. 5.28.

5.7.1 ggH *Results*

The results for the ggH analysis are shown in Fig. 5.29. The observed constraint in the f_{a3}^{ggH} measurement appears to prefer close to the maximum mixture of the CP-odd and CP-even amplitudes with the negative relative sign, with the best fit value at $f_{a3}^{\text{ggH}} = -0.68$. The $\mathcal{D}_{\text{CP}}^{\text{ggH}}$ and $\mathcal{D}_{\text{0-}}^{\text{ggH}}$ distributions in Fig. 5.28 both indicate a preference for about equal contribution of CP-odd and CP-even amplitudes, but are still consistent with the SM expectation of the pure CP-even contribution. This result is statistically consistent with $f_{a3}^{\text{ggH}} = 0$, as expected in the SM, at $1.3\,\sigma$. The

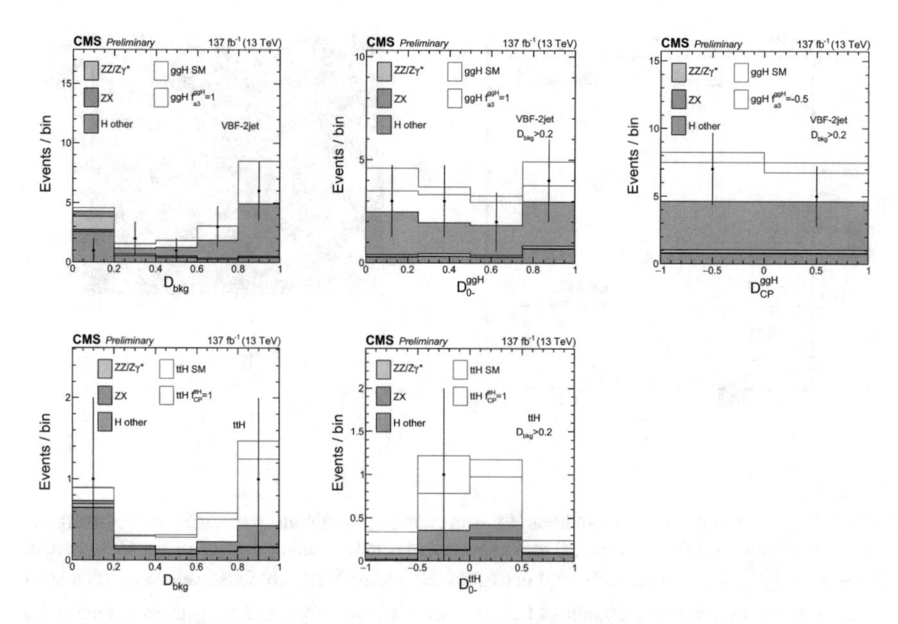

Fig. 5.28 The distributions of events in the observables $\vec{x}$ in the Hff anomalous couplings analysis. The top row shows three of the discriminants used in the VBF-2jet-tagged category: $\mathcal{D}_{\text{bkg}}$, $\mathcal{D}_{0-}$, and $\mathcal{D}_{CP}$. The bottom row shows the observables used in the $\text{t}\bar{\text{t}}\text{H}$ categories: $\mathcal{D}_{\text{bkg}}$ and $\mathcal{D}_{0-}$ [34]

significance of separation of the maximal mixture with the positive relative sign ($f_{a3}^{\text{ggH}} \sim +0.5$) is larger because this scenario would lead to the opposite forward-backward asymmetry in the $\mathcal{D}_{\text{CP}}^{\text{ggH}}$ discriminant distribution shown in Fig. 5.28 for $f_{a3}^{\text{ggH}} \sim -0.5$.

These two parameters f_{a3}^{ggH} and μ_{ggH} are equivalent to the measurement of the CP-even and CP-odd couplings on the production side, while the HVV couplings on the decay side are constrained from the simultaneous measurement of the VBF and VH processes with f_{a3} and μ_V profiled. The c_{gg} and $\tilde{c}_{gg}$ couplings, introduced in Eq. (4.8), can be extracted from the above measurements. We follow the parameterization of the cross section and the total width from [19]. In the total width parameterization, we assume that there are no unobserved or undetected H boson decays. We also assume that fermion couplings Hff are not affected by possible new physics. We allow variation of the HVV and effective Hgg couplings. The former are scaled with the μ_V parameter, and the latter are parameterized with c_{gg} and $\tilde{c}_{gg}$, which describe both SM and BSM contributions to the gluon fusion loop. The small contribution of the $\text{H} \to \gamma\gamma$ and $\text{Z}\gamma$ decays to the total width is assumed to be SM-like. The resulting constraints are shown in Fig. 5.29. The pure signal strength measurement μ_{ggH}, available even without the fit for f_{a3}^{ggH}, provides constraint on $(c_{gg}^2 + \tilde{c}_{gg}^2)$, which is a ring on a two-parameter plane in Fig. 5.29. The measurement of f_{a3}^{ggH} resolves the area within this ring. The H boson width dependence on

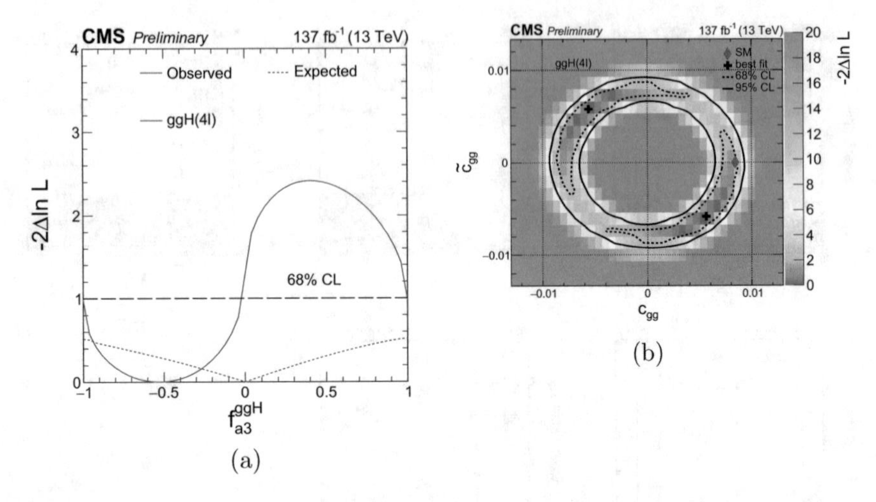

Fig. 5.29 Constraints on the anomalous H boson couplings to gluons in the ggH process using the $\text{H} \to 4\ell$ decay. (**a**) Observed (solid) and expected (dashed) likelihood scans of the CP-sensitive parameter f_{a3}^{ggH}. The dashed horizontal lines show 68 and 95% CL. (**b**) Observed confidence level intervals on the c_{gg} and $\tilde{c}_{gg}$ couplings reinterpreted from the f_{a3}^{ggH} and μ_{ggH} measurement with f_{a3} and μ_V profiled. The dashed and solid lines show the 68% and 95% CL exclusion regions in two dimensions, respectively [34]

c_{gg} and $\tilde{c}_{gg}$ is relatively weak and does not alter this logic considerably. The results are consistent with the SM expectation of $(c_{gg}, \tilde{c}_{gg}) = (0.0084, 0)$ at $1.1\,\sigma$. The correlation between the two parameters is $+0.980$. There is a degeneracy in the measurement between any two points $(c_{gg}, \tilde{c}_{gg})$ and $(-c_{gg}, \tilde{c}_{gg})$, as there is no observable information to resolve this ambiguity.

5.7.2 tt̄H *Results*

Figure 5.30 presents the measurement of anomalous couplings of the H boson to top quarks First, the measurements of f_{CP}^{Htt} from the tt̄H process only are reported. The signal strength $\mu_{\text{t}\bar{\text{t}}\text{H}}$, which is the ratio of the measured cross section of the tt̄H process to that expected in the SM, is profiled when the f_{CP}^{Htt} results are reported. The measured value of $\mu_{\text{t}\bar{\text{t}}\text{H}} = 0.22^{+0.86}_{-0.22}$ is consistent with that reported in [4] without the fit for CP structure of interactions. The correlation between the two parameters is -0.029. The signal strength of the VBF and VH processes μ_V, ggH process μ_{ggH}, and their CP properties f_{a3} and f_{a3}^{ggH} are also profiled when this measurement is performed. This tt̄H analysis is not sensitive to the sign of f_{CP}^{Htt}. However, for later combination with the ggH measurement, presented above, under the assumption of

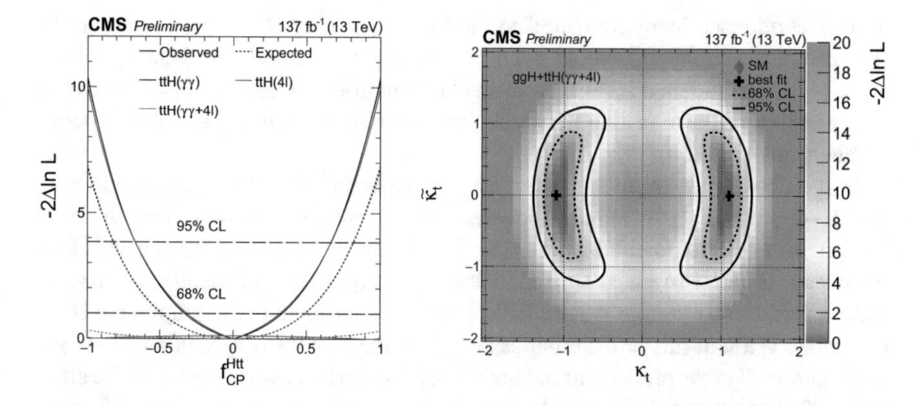

Fig. 5.30 Constraints on the anomalous H boson couplings to top quarks in the $t\bar{t}H$ process using the $H \rightarrow 4\ell$ and $\gamma\gamma$ decays. Left: Observed (solid) and expected (dashed) likelihood scans of f_{CP}^{Htt} in the $t\bar{t}H$ process in the $H \rightarrow 4\ell$ (red), $\gamma\gamma$ (black), and combined (blue) channels, where the combination is done without relating the signal strengths in the two processes. The dashed horizontal lines show 68 and 95% CL. Right: Observed confidence level intervals on the κ_t and $\tilde{\kappa}_t$ couplings reinterpreted from the f_{CP}^{Htt} and $\mu_{t\bar{t}H}$ measurements in the combined fit of the $H \rightarrow 4\ell$ and $\gamma\gamma$ channels, with the signal strength $\mu_{t\bar{t}H}$ in the two channels related through the couplings as discussed in text. The dashed and solid lines show the 68 and 95% CL exclusion regions in two dimensions, respectively

the top quark dominance in the gluon fusion loop, symmetric constraints on f_{CP}^{Htt} are reported.

With just about two signal $t\bar{t}H$ events expected to appear in the fit in the $H \rightarrow 4\ell$ channel under the assumption of the SM cross section, the expected confidence levels of the f_{CP}^{Htt} constraints are low. Nonetheless, the very clean signature in the $H \rightarrow 4\ell$ channel makes any observed event candidate count. The observed best-fit value corresponds to the pure CP-odd Yukawa coupling. This is consistent with the negative value of the $\mathcal{D}_{0-}^{t\bar{t}H}$ discriminant for the one observed signal-like event in Fig. 5.28. However, this result is statistically consistent with the pure CP-even Yukawa coupling expected in the SM at $1.5\,\sigma$.

CMS recently reported the measurement of the f_{CP}^{Htt} parameter in the $t\bar{t}H$ production process with the decay $H \rightarrow \gamma\gamma$ [45]. In that measurement, the signal strength $\mu_{t\bar{t}H}^{\gamma\gamma}$ parameter is profiled, while the signal strengths in other production processes are fixed to the SM expectation. However, there is a very weak correlation of the measurement in the $t\bar{t}H$ process with parameters in the other production mechanisms. We therefore proceed with a combination of the f_{CP}^{Htt} measurements in the $H \rightarrow 4\ell$ and $\gamma\gamma$ channels, where we correlate their common systematic uncertainties, but not the signal strengths of the processes. In particular, we do not relate the $\mu_{t\bar{t}H}$ and $\mu_{t\bar{t}H}^{\gamma\gamma}$ signal strengths because they could be affected differently by the particles appearing in the loops responsible for the $H \rightarrow \gamma\gamma$ decay. The results of this combination are presented in Fig. 5.30. Since the two H boson decay channels have the opposite best-fit values, the combined result has a somewhat smaller confidence level compared to the $H \to \gamma\gamma$ channel alone, excluding the pure pseudoscalar hypothesis at $3.1\,\sigma$. However, the expected exclusion at $2.6\,\sigma$ has a higher confidence level than individual channels. Below we also present an interpretation of these results where the signal strengths in the two H boson decay channels are related.

In the above measurements, the f_{CP}^{Htt} parameter has the same meaning in both the $H \to 4\ell$ and $\gamma\gamma$ channels. However, the signal strength may have different interpretation due to potentially unknown BSM contributions to the loop in the $H \to \gamma\gamma$ decay. In order to make an EFT coupling interpretation of results, we have to make a further assumption that no BSM particles contribute to the loop in the $H \to \gamma\gamma$ decay. Without this or a similar assumption, the signal strength in the $H \to \gamma\gamma$ decay cannot be interpreted without ambiguity. We further re-parameterize the cross section following Ref. [19] with the couplings κ_{t} and $\tilde{\kappa}_{\mathrm{t}}$, and fix $\kappa_{\mathrm{b}} = 1$ and $\tilde{\kappa}_{\mathrm{b}} = 0$. The bottom quark coupling has a very small contribution to the loop in the $H \to \gamma\gamma$ decay, but it has large contribution to the total decay width, where we assume that there are no unobserved or undetected H boson decays. In order to simplify the fit, we do not allow anomalous HVV couplings, and the measurement of the signal strength _muV constrains the contribution of the a_1 coupling in the loop. The f_{a3}^{ggH} and μ_{ggH} parameters are profiled in this fit. The observed confidence level intervals on the κ_{t} and $\tilde{\kappa}_{\mathrm{t}}$ couplings from the combined fit of the $H \to 4\ell$ and $\gamma\gamma$ channels are shown in Fig. 5.30.

5.7.3 Combined Results

The measurement of anomalous couplings of the H boson to top quarks in the ggH process, assuming top quark dominance in the gluon fusion loop, is presented in Fig. 5.31. Similarly to the case of the $H \to \gamma\gamma$ loop discussed above, the cross section of the ggH process, normalized to the SM expectation, is parameterized following Ref. [19] to account for CP-odd Yukawa couplings as follows

$$\frac{\sigma(\mathrm{ggH})}{\sigma_{\mathrm{SM}}} = \kappa_{\mathrm{f}}^2 + 2.38\tilde{\kappa}_{\mathrm{f}}^2 , \tag{5.25}$$

where we set $\kappa_{\mathrm{f}} = \kappa_{\mathrm{t}} = \kappa_{\mathrm{b}}$ and $\tilde{\kappa}_{\mathrm{f}} = \tilde{\kappa}_{\mathrm{t}} = \tilde{\kappa}_{\mathrm{b}}$. Equation (5.25) sets the relationship between the f_{CP}^{Htt} and f_{a3}^{ggH}, reported in Fig. 5.29, according to Eq. (4.6).

Constraints on f_{CP}^{Htt} are also shown with combination of the ggH and $\mathrm{t\bar{t}H}$ processes with $H \to 4\ell$ only and with $H \to \gamma\gamma$ included in the combination, see Fig. 5.31. The gain in this combination of the ggH and $\mathrm{t\bar{t}H}$ processes is beyond the simple addition of the two constraints. While in the ggH and $\mathrm{t\bar{t}H}$ analyses the signal strength of the two processes is independent, these could be related under the assumption of top quark dominance in the loop using Eq. (5.25). As discussed in Sect. 4.4.2, the CP-odd coupling predicts rather different cross sections in the two

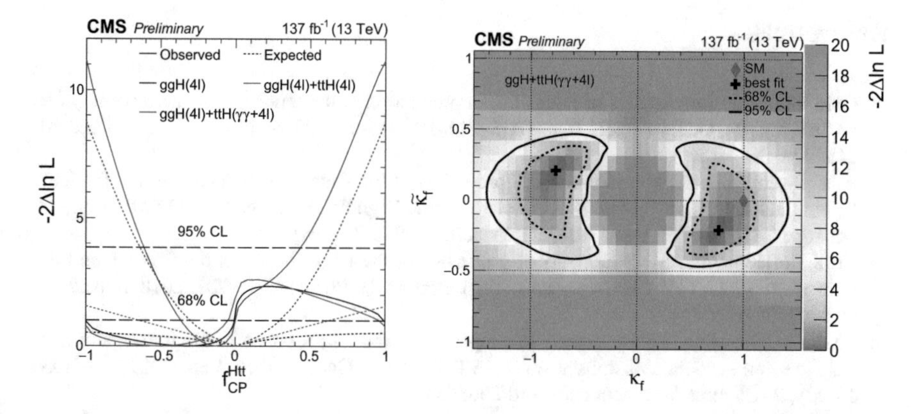

Fig. 5.31 Constraints on the anomalous H boson couplings to top quarks in the $\mathrm{t\bar{t}H}$ and ggH processes combined, assuming top quark dominance in the gluon fusion loop, using the $\mathrm{H} \to 4\ell$ and $\gamma\gamma$ decays. Left: Observed (solid) and expected (dashed) likelihood scans of f_{CP}^{Htt} in the ggH process with $\mathrm{H} \to 4\ell$ (red), $\mathrm{t\bar{t}H}$ and ggH processes combined with $\mathrm{H} \to 4\ell$ (blue), and in the $\mathrm{t\bar{t}H}$ and ggH processes with $\mathrm{H} \to 4\ell$ and the $\mathrm{t\bar{t}H}$ process with $\gamma\gamma$ combined (black). Combination is done by relating the signal strengths in the three processes through the couplings in the loops in both production and decay, as discussed in the text. The dashed horizontal lines show 68 and 95% CL exclusion. Right: Observed confidence level intervals on the κ_t and $\tilde{\kappa}_\mathrm{t}$ couplings reinterpreted from the f_{CP}^{Htt} and signal strength measurements in the fit corresponding to the full combination of $\mathrm{t\bar{t}H}$ and ggH processes and the $\mathrm{H} \to 4\ell$ and $\gamma\gamma$ channels in the left plot. The dashed and solid lines show the 68 and 95% CL exclusion regions in two dimensions, respectively

processes: $\sigma(\tilde{\kappa}_f = 1)/\sigma(\kappa_f = 1)$ is 2.38 in the gluon fusion process dominated by the top quark loop and 0.391 in the $\mathrm{t\bar{t}H}$ process. This means that the ratio differs by a factor of 6.09 for $f_{CP}^{\mathrm{Htt}} = 1$ when compared to SM ($f_{CP}^{\mathrm{Htt}} = 0$). This correlation enhances the sensitivity in the f_{CP}^{Htt} measurement.

The combination of the $\mathrm{H} \to 4\ell$ and $\gamma\gamma$ channels in combination of the ggH and $\mathrm{t\bar{t}H}$ processes proceed in a manner similar. In particular, we do not allow anomalous HVV couplings, and the measurement of the signal strength μ_V constrains the contribution of the a_1 coupling in the $\mathrm{H} \to \gamma\gamma$ loop. The full combination of the f_{CP}^{Htt} results is also shown in Fig. 5.31.

Finally, the re-interpretation of the f_{CP}^{Htt} and signal strength measurements in terms of constraints on κ_f and $\tilde{\kappa}_\mathrm{f}$ shown in Fig. 5.31. In this fit, it is assumed that $\kappa_\mathrm{f} = \kappa_\mathrm{t} = \kappa_\mathrm{b} = \kappa_\mathrm{c} = \kappa_\mu$ and $\tilde{\kappa}_\mathrm{f} = \tilde{\kappa}_\mathrm{t} = \tilde{\kappa}_\mathrm{b} = \tilde{\kappa}_\mathrm{c} = \tilde{\kappa}_\mu$ in the fermion coupling contribution to the production processes and in the decay width parameterization [19]. The measurement of the signal strength μ_V constrains the contributions of the a_1 coupling and anomalous HVV couplings are not allowed. It is assumed that there are no unobserved or undetected H boson decays.

References

1. CMS Collaboration, Studies of Higgs boson production in the four-lepton final state at $\sqrt{s} =$ 13 TeV. CERN, Geneva, Tech. Rep. CMS-PAS-HIG-15-004, 2016. https://cds.cern.ch/record/2139978
2. A.M. Sirunyan et al., Measurements of properties of the Higgs boson decaying into the four-lepton final state in **pp** collisions at $\sqrt{s} = 13$ TeV. J. High Energy Phys. **11**, 047 (2017). https://doi.org/10.1007/JHEP11(2017)047. arXiv: 1706.09936 [hep-ex]
3. CMS Collaboration, Measurements of properties of the Higgs boson in the four-lepton final state at $\sqrt{s} = 13$ TeV. CERN, Geneva, Tech. Rep. CMS-PAS-HIG-18-001, 2018. https://cds.cern.ch/record/2621419
4. CMS Collaboration, Measurements of properties of the Higgs boson in the four-lepton final state in proton-proton collisions at $\sqrt{s} = 13$ TeV. CERN, Geneva, Tech. Rep. CMS-PAS-HIG-19-001, 2019. https://cds.cern.ch/record/2668684
5. S. Chatrchyan et al., On the mass and spin-parity of the Higgs boson candidate via its decays to Z boson pairs. Phys. Rev. Lett. **110**, 081803 (2013) . https://doi.org/10.1103/PhysRevLett.110.081803. arXiv: 1212.6639 [hep-ex]
6. S. Chatrchyan et al., Measurement of the properties of a Higgs boson in the four-lepton final state. Phys. Rev. D **89**, 092007 (2014). https://doi.org/10.1103/PhysRevD.89.092007. arXiv: 1312.5353 [hep-ex]
7. V. Khachatryan et al., Constraints on the spin-parity and anomalous HVV couplings of the Higgs boson in proton collisions at 7 and 8 TeV. Phys. Rev. D **92**, 012004 (2015). https://doi.org/10.1103/PhysRevD.92.012004. arXiv: 1411.3441 [hep-ex]
8. G. Aad et al., Evidence for the spin-0 nature of the Higgs boson using ATLAS data. Phys. Lett. B **726**, 120 (2013). https://doi.org/10.1016/j.physletb.2013.08.026. arXiv: 1307.1432 [hep-ex]
9. G. Aad et al., Study of the spin and parity of the Higgs boson in diboson decays with the ATLAS detector. Eur. Phys. J. C **75**, 476 (2015). https://doi.org/10.1140/epjc/s10052-015-3685-1. arXiv: 1506.05669 [hep-ex]
10. S. Frixione, P. Nason, C. Oleari, Matching NLO QCD computations with parton shower simulations: the POWHEG method. J. High Energy Phys. **11**, 070 (2007). https://doi.org/10.1088/1126-6708/2007/11/070. arXiv: 0709.2092 [hep-ph]
11. E. Bagnaschi, G. Degrassi, P. Slavich, A. Vicini, Higgs production via gluon fusion in the POWHEG approach in the SM and in the MSSM. J. High Energy Phys. **2**, 088 (2012). https://doi.org/10.1007/JHEP02(2012)088. arXiv: 1111.2854 [hep-ph]
12. P. Nason, C. Oleari, NLO Higgs boson production via vector-boson fusion matched with shower in POWHEG. J. High Energy Phys. **2**, 037 (2010). https://doi.org/10.1007/JHEP02(2010)037. arXiv: 0911.5299 [hep-ph]
13. G. Luisoni, P. Nason, C. Oleari, F. Tramontano, $HW^{\pm}/HZ + 0$ and 1 jet at NLO with the POWHEG BOX interfaced to GoSam and their merging within MiNLO. J. High Energy Phys. **10**, 083 (2013). https://doi.org/10.1007/JHEP10(2013)083. arXiv: 1306.2542 [hep-ph]
14. H.B. Hartanto, B. Jager, L. Reina, D.Wackeroth, Higgs boson production in association with top quarks in the POWHEG BOX. Phys. Rev. D **91**, 094003 (2015). https://doi.org/10.1103/PhysRevD.91.094003. arXiv: 1501.04498 [hep-ph]
15. Y. Gao, A.V. Gritsan, Z. Guo, K. Melnikov, M. Schulze, N.V. Tran, Spin determination of single-produced resonances at hadron colliders. Phys. Rev. D **81**, 075022 (2010). https://doi.org/10.1103/PhysRevD.81.075022. arXiv: 1001.3396 [hep-ph]
16. S. Bolognesi, Y. Gao, A.V. Gritsan, K. Melnikov, M. Schulze, N.V. Tran, A. Whitbeck, Spin and parity of a single-produced resonance at the LHC. Phys. Rev. D **86**, 095031 (2012). https://doi.org/10.1103/PhysRevD.86.095031. arXiv: 1208.4018 [hep-ph]
17. I. Anderson, S. Bolognesi, F. Caola, Y. Gao, A.V. Gritsan, C.B. Martin, K. Melnikov, M. Schulze, N.V. Tran, A. Whitbeck, Y. Zhou, Constraining anomalous HVV interactions at proton and lepton colliders. Phys. Rev. D **89**, 035007 (2014). https://doi.org/10.1103/PhysRevD.89.035007. arXiv: 1309.4819 [hep-ph]

18. A.V. Gritsan, R. Röntsch, M. Schulze, M. Xiao, Constraining anomalous Higgs boson couplings to the heavy flavor fermions using matrix element techniques. Phys. Rev. D **94**, 055023 (2016). https://doi.org/10.1103/PhysRevD.94.055023. arXiv: 1606.03107 [hep-ph]
19. A. V. Gritsan, J. Roskes, U. Sarica, M. Schulze, M. Xiao, and Y. Zhou, "New features in the JHU generator framework: constraining Higgs boson properties from on-shell and off-shell production," *Phys. Rev. D*, vol. 102, no. 5, p. 056022, 2020. https://doi.org/10.1103/PhysRevD.102.056022.arXiv:2002.09888 [hep-ph]
20. J.M. Campbell, R.K. Ellis, MCFM for the Tevatron and the LHC. Nucl. Phys. Proc. Suppl. **205–206**, 10 (2010). https://doi.org/10.1016/j.nuclphysbps.2010.08.011. arXiv: 1007.3492 [hep-ph]
21. R.J. Barlow, Extended maximum likelihood. Nucl. Instrum. Meth. A **297**, 496 (1990). https://doi.org/10.1016/0168-9002(90)91334-8
22. V. Khachatryan et al., Combined search for anomalous pseudoscalar HVV couplings in VH(H $\rightarrow$ b$\bar{\text{b}}$) production and H $\rightarrow$ VV decay. Phys. Lett. B **759**, 672 (2016). https://doi.org/10.1016/j.physletb.2016.06.004. arXiv: 1602.04305 [hep-ex]
23. G. Aad et al., Test of CP invariance in vector-boson fusion production of the Higgs boson using the optimal observable method in the ditau decay channel with the ATLAS detector. Eur. Phys. J. C **76**, 658 (2016). https://doi.org/10.1140/epjc/s10052-016-4499-5. arXiv: 1602.04516 [hep-ex]
24. V. Khachatryan et al., Limits on the Higgs boson lifetime and width from its decay to four charged leptons. Phys. Rev. D **92**, 072010 (2015). https://doi.org/10.1103/PhysRevD.92.072010. arXiv: 1507.06656 [hep-ex]
25. G. Aad et al., Constraints on the off-shell Higgs boson signal strength in the high-mass ZZ and WW final states with the ATLAS detector. Eur. Phys. J. C **75**, 335 (2015). https://doi.org/10.1140/epjc/s10052-015-3542-2. arXiv: 1503.01060 [hep-ex]
26. A.M. Sirunyan et al., Constraints on anomalous Higgs boson couplings using production and decay information in the four-lepton final state. Phys. Lett. B **775**, 1 (2017). https://doi.org/10.1016/j.physletb.2017.10.021. arXiv: 1707.00541 [hep-ex]
27. A.M. Sirunyan et al., Measurements of the Higgs boson width and anomalous *HVV* couplings from on-shell and off-shell production in the four-lepton final state. Phys. Rev. D vol. **99**(11), 112003 (2019). https://doi.org/10.1103/PhysRevD.99.112003. arXiv: 1901.00174 [hep-ex]
28. A.M. Sirunyan et al., Search for a new scalar resonance decaying to a pair of Z bosons in proton-proton collisions at $\sqrt{s} = 13$ TeV. J. High Energy Phys. **6**, 127 (2018). [Erratum: JHEP **3**, 128 (2019)]. https://doi.org/10.1007/JHEP06(2018)127. arXiv: 1804.01939 [hep-ex]
29. A.M. Sirunyan et al., Constraints on anomalous *HVV* couplings from the production of Higgs bosons decaying to τ lepton pairs. Phys. Rev. D **100**(11), 112002 (2019). https://doi.org/10.1103/PhysRevD.100.112002. arXiv: 1903.06973 [hep-ex]
30. A.M. Sirunyan et al., Observation of the Higgs boson decay to a pair of τ leptons with the CMS detector. Phys. Lett. **B779**, 283–316 (2018). https://doi.org/10.1016/j.physletb.2018.02.004. arXiv: 1708.00373 [hep-ex]
31. V. Khachatryan et al., Reconstruction and identification of τ lepton decays to hadrons and ν_τ at CMS. JINST **11**, P01019 (2016). https://doi.org/10.1088/1748-0221/11/01/P01019. arXiv: 1510.07488 [physics.ins-det]
32. A.M. Sirunyan et al., Performance of reconstruction and identification of τ leptons decaying to hadrons and ν_τ in pp collisions at $\sqrt{s} = 13$ TeV. JINST **13**, P10005 (2018). https://doi.org/10.1088/1748-0221/13/10/P10005. arXiv: 1809.02816 [hep-ex]
33. L. Bianchini, J. Conway, E.K. Friis, C. Veelken, Reconstruction of the Higgs mass in $H \rightarrow \tau\tau$ events by dynamical likelihood techniques. J. Phys. Conf. Ser. **513**, 022035 (2014). https://doi.org/10.1088/1742-6596/513/2/022035.
34. CMS collaboration, Constraints on anomalous Higgs boson couplings to vector bosons and fermions in production and decay $H \rightarrow 4l$channel. Report No.: CMS-PAS-HIG-19-009, to be submitted to *Phys. Rev. D*. http://cds.cern.ch/record/2725543
35. H. Stetter, *Numerical Polynomial Algebra* (Society for Industrial and Applied Mathematics, 2004). https://doi.org/10.1137/1.9780898717976

36. B. Buchberger, A theoretical basis for the reduction of polynomials to canonical forms. SIGSAM Bull. **10**(3), 19–29 (1976). ISSN: 0163-5824. https://doi.org/10.1145/1088216.1088219
37. A. Kondratyev, H.J. Stetter, F. Winkler, Numerical Computation of Gröbner Bases (Citeseer, 2004), pp. 295–306
38. T. Chen, T.-Y. Li, Homotopy continuation method for solving systems of nonlinear and polynomial equations. Commun. Inf. Syst. **15**(2), 119–307 (2015). ISSN: 15267555, 21634548. https://doi.org/10.4310/CIS.2015.v15.n2.a1
39. T. Chen, T.-L. Lee, T.-Y. Li, Hom4PS-3: A parallel numerical solver for systems of polynomial equations based on polyhedral homotopy continuation methods, in *Mathematical Software—ICMS 2014*, ed. by H. Hong, C. Yap, Lecture Notes in Computer Science, vol. 8592 (Springer, Berlin, 2014), pp. 183–190. ISBN: 978-3-662-44199-2. https://doi.org/10.1007/978-3-662-44199-2_30
40. T. Chen, T.-L. Lee, and T.-Y. Li, Mixed cell computation in Hom4PS-3. J. Symb. Comput. **79**(Part 3), 516–534 (2017), ISSN: 0747-7171. https://doi.org/10.1016/j.jsc.2016.07.017
41. A.S. Nemirovsky, D.B. Yudin, *Problem Complexity and Method Efficiency in Optimization* (Wiley, London, 1983)
42. S. Diamond, S. Boyd, CVXPY: A Python-embedded modeling language for convex optimization. J. Mach. Learn. Res. **17**(83), 1–5 (2016)
43. A. Agrawal, R. Verschueren, S. Diamond, S. Boyd, A rewriting system for convex optimization problems. J. Control Decis. **5**(1), 42–60 (2018)
44. Gurobi Optimization, LLC, *Gurobi Optimizer Reference Manual* (2018). http://www.gurobi.com
45. A.M. Sirunyan et al., Measurements of $t\bar{t}H$ production and the CP structure of the Yukawa interaction between the Higgs boson and top quark in the diphoton decay channel (2020). arXiv: 2003.10866 [hep-ex]

Chapter 6
Conclusions and Future Directions

This thesis presents several analyses of Higgs boson anomalous couplings. Over time, these analyses have improved not only by increasing the amount of data, but also by developing new methods to increase the sensitivity and run more and more complicated configurations. We have measured anomalous HVV couplings from both decay and from VBF and VH production in a model independent way that floats multiple anomalous couplings at the same time, excluded the pure pseudoscalar hypothesis for Hff couplings at 68% confidence level, and found the width of the Higgs boson to be nonzero and not more than 2.2 times its Standard Model value at 95% confidence level.

What remains to be done? One obvious next step is to fully utilize all of the data that we have. Analyses using production information can benefit by looking at multiple channels at once. The analysis described in Sect. 5.5 is a start in this direction, and similar analyses can be performed in $\mathrm{H} \to \gamma\gamma$ and $\mathrm{H} \to \mathrm{b\bar{b}}$. The analysis can also be done in $\mathrm{H} \to \mathrm{WW}$, which is more similar to $\mathrm{H} \to \mathrm{ZZ}$ because it has an HVV vertex on the decay side as well. The Hff analysis can also be extended to these channels and may be able to reach 95% confidence level exclusion, even with the current data, through a combination of multiple channels.

It is also possible to look at the phase of various couplings to see if it matches the Standard Model expectation. The offshell region includes interference between signal and background. Interference in offshell gluon fusion production is sensitive to the $\phi_{\mathrm{Htt}} + \phi_{\mathrm{HVV}}$, while interference in VBF and VH production is sensitive to $2\phi_{\mathrm{HVV}}$. Other rare processes, such as $\mathrm{gg} \to \mathrm{ZH}$ and tqH, which have not been observed yet, provide a unique handle on the interference between HVV and Hff couplings and can be used to measure $\phi_{\mathrm{HVV}} - \phi_{\mathrm{Htt}}$. These phases might be sensitive to BSM effects.

Run 3 of the LHC will more than double the current dataset, and the HL-LHC, to be installed starting in 2023, will increase that by a further order of magnitude. At that point, it is natural to wonder if it is still interesting to search for anomalous couplings in the Higgs boson decay. After all, VBF and VH dominate for small

J. Roskes, *A Boson Learned from its Context, and a Boson Learned from its End*, Springer Theses, https://doi.org/10.1007/978-3-030-58011-7_6

anomalous couplings, and with the increased dataset they will be able to measure those small anomalous couplings at high confidence level. However, it is possible to consider further q^2 dependence of anomalous couplings. If anomalous couplings are produced through loops involving BSM particles (or, for that matter, SM particles like the top quark), their effect will change on energy scales around the mass of those particles. If any anomalous couplings are observed, their q^2 dependence will help to understand what new physics is causing them. The results we have already seen in Fig. 5.23, with the decay information minimizing away from $\vec{f_{ai}} = 0$ when multiple anomalous couplings are floated, may even be a hint towards this kind of q^2 dependence.

Figure 6.1 [1] shows scans of f_{ai} parameters using the MELA approach; a Simplified Template Cross Section (STXS) approach [2]; and using decay information only. These scans are projected to the full planned HL-LHC luminosity of 3000 fb^{-1}. The red curve excludes much smaller values of the anomalous couplings than the green one, but the two curves probe different q^2 regions and both remain interesting—especially if the Run 3 data confirms the minimum away from 0 seen in the current analysis.

In the absence of q^2 dependence, the fits, dominated by production information, can be reinterpreted as constraints on the EFT parameters defined in Sect. 4.4.3. In these fits, the anomalous couplings are related by the $SU(2) \times U(1)$ in Eqs. (4.9)–(4.13), with $a_{2,3}^{Z\gamma,\gamma\gamma}$ fixed to 0. Three anomalous couplings are floated, corresponding to a_2, a_3, and Λ_1, with $\Lambda_1^{Z\gamma}$ defined by a linear combination of those couplings. Expected one- and two-dimensional constraints [1] on the EFT parameters are shown in Figs. 6.2 and 6.3.

All of the methods described here can be naturally and directly applied to other Standard Model processes. For example, vector boson scattering is similar to vector boson fusion, but instead of a Higgs boson, it include various triple or quartic gauge boson couplings. The final state for this process is two vector bosons, exactly the same as we investigate here, and matrix element discriminants can be naturally used in exactly the same way to probe those couplings. Anomalous couplings in those processes can be as useful in probing new physics as anomalous couplings of the Higgs boson.

We know physics beyond the Standard Model is out there. It exists in the form of gravity, dark matter, dark energy, and several other puzzles. The challenge is to find where that new physics is hiding in Standard Model language. These results have eliminated several of those possible hiding places and provided methods for illuminating more.

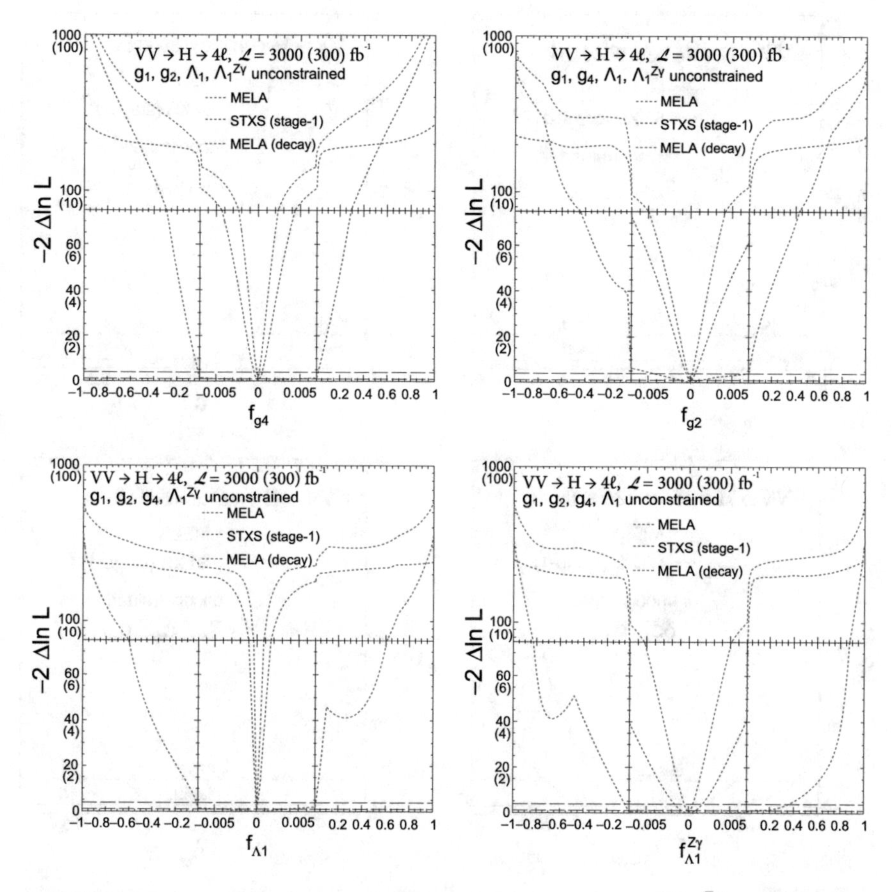

Fig. 6.1 Expected constraints from a simultaneous fit of f_{g4}, f_{g2}, $f_{\Lambda 1}$, and $f_{\Lambda 1}^{Z\gamma}$ using associated production and H $\rightarrow$ 4ℓ decay with 3000 (300) fb^{-1} data. Three analysis scenarios are shown: using MELA observables with production and decay (or decay only) information, and using STXS binning. The dashed horizontal lines show the 68 and 95% CL regions [1]

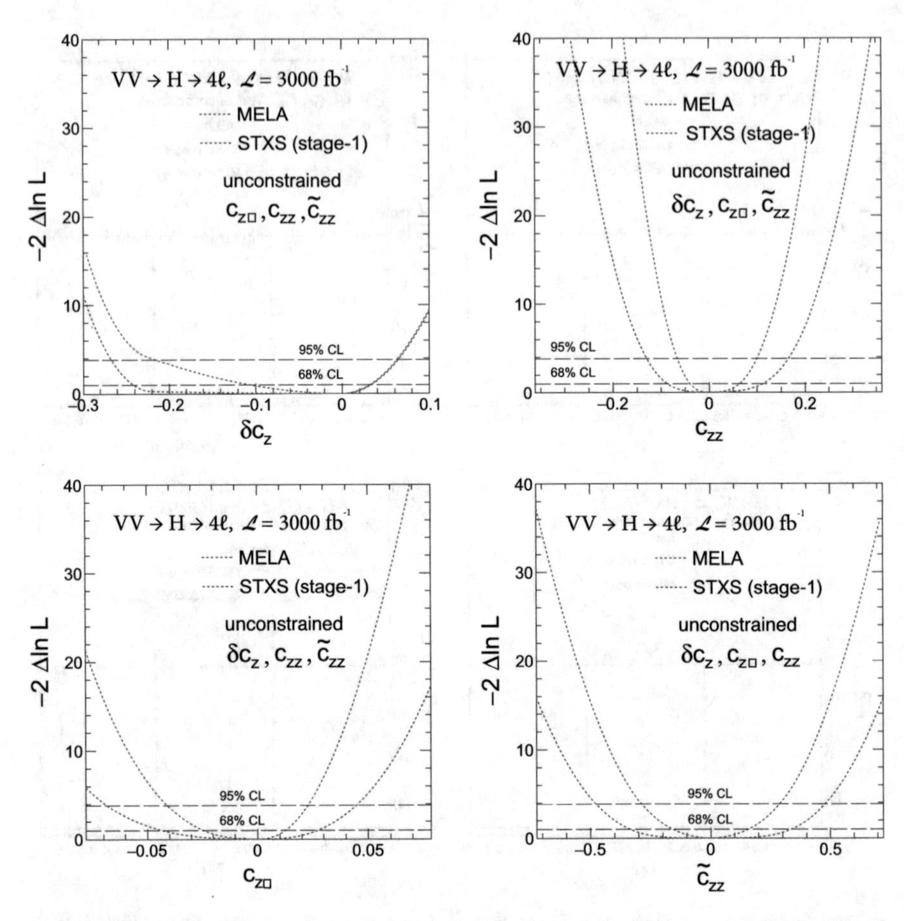

Fig. 6.2 Expected constraints from a simultaneous fit of δc_z, c_{zz}, $c_{z\Box}$, and $\tilde{c}_{zz}$ using associated production and $\mathrm{H} \to 4\ell$ decay with $3000\,\mathrm{fb}^{-1}$ data. The constraints on each parameter are shown with the other parameters describing the HVV and Hgg couplings profiled. Two analysis scenarios are shown: using MELA observables and using STXS binning [1]

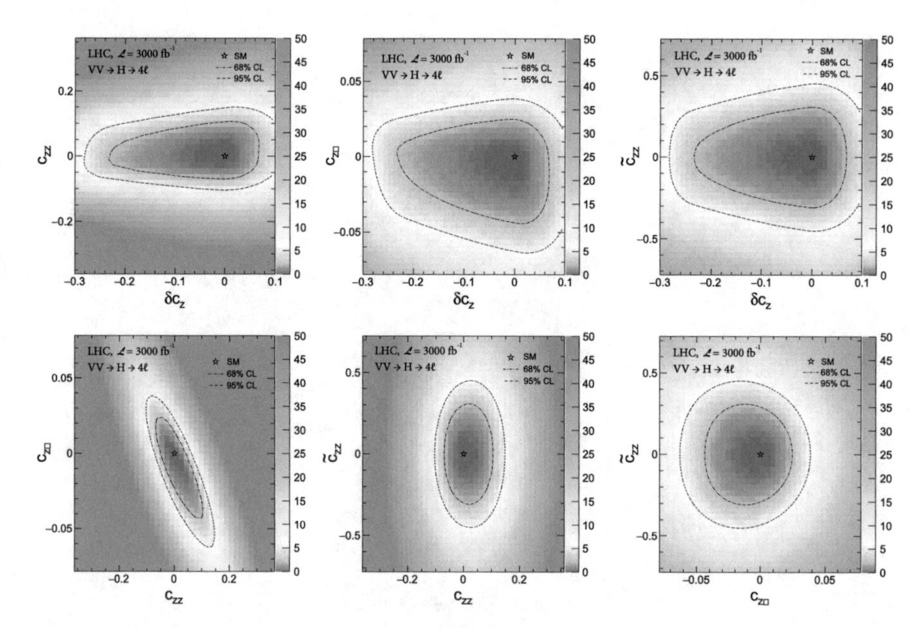

Fig. 6.3 Expected two-dimensional constraints from a simultaneous fit of δc_z, c_{zz}, $c_{z\Box}$, and $\tilde{c}_{zz}$ as shown in Fig. 6.2 for the MELA observables. The constraints on each parameter are shown with the other parameters describing the HVV and Hgg couplings profiled. Top-left: $(\delta c_z, c_{zz})$; top-middle: $(\delta c_z, c_{z\Box})$; top-right: $(\delta c_z, \tilde{c}_{zz})$; bottom-left: $(c_{zz}, c_{z\Box})$; bottom-middle: $(c_{zz}, \tilde{c}_{zz})$; bottom-right: $(c_{z\Box}, \tilde{c}_{zz})$ [1]

References

1. A. V. Gritsan, J. Roskes, U. Sarica, M. Schulze, M. Xiao, and Y. Zhou, "New features in the JHU generator framework: constraining Higgs boson properties from on-shell and off-shell production," *Phys. Rev. D*, vol. 102, no. 5, p. 056022, 2020. https://doi.org/10.1103/PhysRevD.102.056022.arXiv:2002.09888 [hep-ph]
2. CMS Collaboration, Measurements of properties of the Higgs boson in the four-lepton final state in proton-proton collisions at $\sqrt{s}$ = 13 TeV. Technical Report CMS-PAS-HIG-19-001, CERN, Geneva (2019) [Online]. Available: https://cds.cern.ch/record/2668684

Vita

Jeffrey Roskes

Email: jroskes1@jhu.edu *Inspire HEP ID:* 1319409

Professional Experience

Johns Hopkins University, Baltimore, MD
Institute for Data Intensive Engineering and Science
Assistant Research Scientist, 2020–present
  Apply data science techniques to precise medical imaging
  Collect and calibrate high quality, precise data and build a large dataset
  Run statistical analyses to improve patient treatment

J. Roskes, *A Boson Learned from its Context, and a Boson Learned from its End*, Springer Theses, https://doi.org/10.1007/978-3-030-58011-7

Education

Johns Hopkins University, Baltimore, MD
Ph.D., Physics, 2019
Dissertation: *A boson learned from its context, and a boson learned from its end*
Advisor: Prof. Andrei Gritsan
Johns Hopkins University, Baltimore, MD
M.A., Physics, 2015
Johns Hopkins University, Baltimore, MD
B.S., Physics (Departmental Honors and General University Honors), 2014
Second major, Mathematics, 2014
Cumulative GPA: 3.94/4.00

Graduate Research Assistant
William Gardner Fellow, 2016
Donald E. Kerr Sr. and Barbara Kerr Stanley Predoctoral Fellow, 2014–2015
Johns Hopkins University
August 2014–October 2019, *with Prof. Andrei Gritsan*
CMS Higgs boson analysis
CMS tracker alignment
Undergraduate Research Assistant
Johns Hopkins University
Summer 2012–Summer 2014, *with Prof. Andrei Gritsan*
CMS tracker alignment

Leadership Experience

Convener, CMS tracker alignment group, 2018
Coordinated the calibration of the CMS tracker, measuring the positions of over 17000 modules
Led weekly meetings to discuss strategies and results
Facilitated communication between the tracker alignment group and related groups within CMS
Monte Carlo contact, CMS HZZ group, 2017–2019
Prepared Monte Carlo simulation requests for analyses within the group
Facilitated communication between the HZZ group and the Monte Carlo generation group

---

Publications

Computational Biology

20. S. Berry, N. Giraldo, B. Green, *et al.*, "Analysis of multispectral imaging with the AstroPath platform informs efficacy of PD-1 blockade," submitted to *Science*

Particle Physics

As a CMS collaboration member, I am a co-author of 524 CMS publications. See Inspire HEP for the full list.

19. CMS collaboration, "Alignment Strategies and Performance of CMS silicon tracker during LHC Run-2 *(working title)*," to be submitted to *Nuclear Instruments and Methods in Physics Research A*
18. CMS collaboration, Constraints on anomalous Higgs boson couplings to vector bosons and fermions in production and decay $H \rightarrow 4l$channel. Report No.: CMS-PAS-HIG-19-009, to be submitted to *Phys. Rev. D.* http://cds.cern.ch/record/2725543
17. A. V. Gritsan, J. Roskes, U. Sarica, M. Schulze, M. Xiao, and Y. Zhou, "New features in the JHU generator framework: constraining Higgs boson properties from on-shell and off-shell production," *Phys. Rev. D*, vol. 102, no. 5, p. 056022, 2020. https://doi.org/10.1103/PhysRevD.102.056022.arXiv: 2002.09888 [hep-ph]
16. M. Cepeda *et al.*, "Report from Working Group 2: Higgs Physics at the HLLHC and HE-LHC," in *Report on the Physics at the HL-LHC,and Perspectives for the HE-LHC*, A. Dainese, M. Mangano, A. B. Meyer, *et al.*, Eds. Dec. 2019, vol. 7, pp. 221–584. https://doi.org/10.23731/CYRM-2019-007.221. arXiv: 1902.00134 [hep-ph]
15. CMS Collaboration, "Measurements of properties of the Higgs boson in the four-lepton final state in proton-proton collisions at $\sqrt{s} = 13$ TeV," CERN, Geneva, Tech. Rep. CMS-PAS-HIG-19-001, 2019. [Online]. Available: https://cds.cern.ch/record/2668684
14. A. M. Sirunyan *et al.*, "Constraints on anomalous *HVV* couplings from the production of Higgs bosons decaying to τ lepton pairs," *Phys. Rev. D*, vol. 100, no. 11, p. 112 002, 2019. https://doi.org/10.1103/PhysRevD.100.112002. arXiv: 1903.06973 [hep-ex]

13. A. M. Sirunyan *et al.*, “Measurements of the Higgs boson width and anomalous *HVV* couplings from on-shell and off-shell production in the four-lepton final state,” *Phys. Rev. D*, vol. 99, no. 11, p. 112 003, 2019. https://doi.org/10.1103/PhysRevD.99.112003. arXiv: 1901.00174 [hep-ex]
12. CMS Collaboration, “CMS Tracker Alignment Performance Results 2018,” Nov. 2018. [Online]. Available: https://cds.cern.ch/record/2650977
11. “Measurements of properties of the Higgs boson in the four-lepton final state at $\sqrt{s}$ = 13 TeV,” CERN, Geneva, Tech. Rep. CMS-PAS-HIG-18-001, 2018. [Online]. Available: https://cds.cern.ch/record/2621419
10. A. M. Sirunyan *et al.*, “Search for a new scalar resonance decaying to a pair of Z bosons in proton-proton collisions atÂ $\sqrt{s}$ = 13 TeV,” *JHEP*, vol. 06, p. 127, 2018, [Erratum: JHEP 03, 128 (2019)]. https://doi.org/10.1007/JHEP06(2018)127. arXiv: 1804.01939 [hep-ex]
9. CMS Collaboration, “Tracker Alignment Performance Plots after Commissioning,” Dec. 2017. [Online]. Available: http://cds.cern.ch/record/2297526
8. CMS Collaboration, “CMS Tracker Alignment Performance Results Start-Up 2017,” Dec. 2017. [Online]. Available: http://cds.cern.ch/record/2297528
7. A. M. Sirunyan *et al.*, “Constraints on anomalous Higgs boson couplings using production and decay information in the four-lepton final state,” *Phys. Lett. B*, vol. 775, p. 1, 2017. https://doi.org/10.1016/j.physletb.2017.10.021. arXiv: 1707.00541 [hep-ex]
6. A. M. Sirunyan *et al.*, “Measurements of properties of the Higgs boson decaying into the four-lepton final state in pp collisions atÂ $\sqrt{s}$ = 13 TeV,” *JHEP*, vol. 11, p. 047, 2017. https://doi.org/10.1007/JHEP11(2017)047. arXiv: 1706.09936 [hep-ex]
5. D. de Florian, C. Grojean, *et al.*, “Handbook of LHC Higgs cross sections: 4. deciphering the nature of the Higgs sector,” 2016. https://doi.org/10.23731/CYRM-2017-002. arXiv: 1610.07922
4. CMS Collaboration, “CMS Tracker Alignment Performance Results 2016,” Jun. 2017. [Online]. Available: http://cds.cern.ch/record/2273267
3. V. Khachatryan *et al.*, “Limits on the Higgs boson lifetime and width from its decay to four charged leptons,” *Phys. Rev. D*, vol. 92, p. 072 010, 2015. https://doi.org/10.1103/PhysRevD.92.072010. arXiv: 1507.06656 [hep-ex]
2. CMS Collaboration, “Alignment of the CMS Tracking-Detector with First 2015 Cosmic-Ray and Collision Data,” Aug. 2015. [Online]. Available: http://cds.cern.ch/record/2041841
1. V. Khachatryan *et al.*, “Constraints on the spin-parity and anomalous HVV couplings of the Higgs boson in proton collisions at 7 and 8 TeV,” *Phys. Rev. D*, vol. 92, p. 012 004, 2015. https://doi.org/10.1103/PhysRevD.92.012004. arXiv: 1411.3441 [hep-ex]

JHU Generator and MELA package, http://spin.pha.jhu.edu/

The JHU Generator simulates a wide variety of processes involving the Higgs boson or a new spin-0, 1, or 2 resonance. The MELA (Matrix Element Likelihood

Approach) package calculates probabilities for these events, used to reweight the simulated events and to construct discriminants to be applied to data and distinguish between hypotheses.

Presentations

9. J. Roskes, "JHU generator framework: New features for Higgs boson studies," presented at the Phenomenology 2020 Symposium (Pittsburgh, PA; virtual due to COVID-19), May 2020, https://indico.cern.ch/event/858682/contributions/3837206/
8. J. Roskes, "Higgs boson anomalous couplings and width at CMS," presented at the JHU HEP/Cosmology Seminar (Baltimore, MD), Feb. 2019, http://physics-astronomy.jhu.edu/events/particle-physicsseminars/
7. J. Roskes, "Measuring the Higgs boson width and anomalous HVV couplings in production and decay with CMS," presented at the Lake Louise Winter Institute 2019 (Lake Louise, AB), Feb. 2019, https://indico.cern.ch/event/760557/contributions/3262341/
6. J. Roskes, "Recent developments in H $\rightarrow$ $4l$ analyses on CMS," presented at the Higgs Couplings 2016 Workshop (Menlo Park, CA), Nov. 2016, https://indico.cern.ch/event/477407/timetable/
5. J. Roskes, "Recent developments in H $\rightarrow$ $4l$ analyses on CMS," presented at the JHU HEP/Cosmology Seminar (Baltimore, MD), Nov. 2016, http://physics-astronomy.jhu.edu/events/particle-physicsseminars/
4. J. Roskes, "CMS tracker alignment and status of the CMS tracker in Run2," presented at the JHU HEP/Cosmology Seminar (Baltimore, MD), Sep. 2015, http://physics-astronomy.jhu.edu/events/particle-physics-seminars/
3. J. Roskes, "Tools for the higgs boson *CP* studies: JHUGen and MELA," presented at the 2015 Meeting of the American Physical Society Division of Particles and Fields (Ann Arbor, MI), Aug. 2015, https://indico.cern.ch/event/361123/call-for-abstracts/307/
2. J. Roskes, "CMS silicon tracker alignment: First Run2 results," presented at the 2015 Meeting of the American Physical Society Division of Particles and Fields (Ann Arbor, MI), Aug. 2015, https://indico.cern.ch/event/361123/call-for-abstracts/77/
1. J. Roskes, "Validation of the Higgs boson spin-parity analysis with Z $\rightarrow$ $4l$ data," presented at the American Physical Society April Meeting 2015 (Baltimore, MD), Apr. 2015, http://meetings.aps.org/Meeting/APR15/Session/X16.8

Teaching and Mentoring

Mentoring

During my time in graduate school, I mentored 9 undergraduate and 2 younger graduate students in various aspects of CMS research. In particular, three of the undergraduate students are co-authors of the CMS paper documenting tracker alignment in Run 2, number 19 on the publication list above, which includes their work.

Teaching Assistant

Introduction to Practical Data Science: Beautiful Data (Spring 2019)

Introduction to Practical Data Science: Beautiful Data (Fall 2017)

General Physics II for Biological Science Majors, Johns Hopkins University (Spring 2015)

General Physics Laboratory II, Johns Hopkins University (Spring 2015)

General Physics I for Biological Science Majors, Johns Hopkins University (Fall 2014)

General Physics Laboratory I, Johns Hopkins University (Fall 2014)

Outreach

Volunteer, Physics Fair, Johns Hopkins University, 2015–2018

Prepared demonstrations and exciting videos showcasing physical phenomena

Volunteer, LHC exhibit, USA Science and Engineering Festival, Washington, DC, 2014, 2016, and 2018

Prepared and set up cosmic ray detector and other demonstrations

Explained the concept of particle accelerators and detectors to all visitors, especially elementary school children

Counselor, Stevenson University Science Camp, Summers 2008–2014

Delivered the scientific content of the camp

Encouraged middle-school students in scientific activities

Member, Sheridan Libraries Student Advisory Committee, Johns Hopkins University, 2011–2014

Recommended ways to improve library services

Printed in the United States
by Baker & Taylor Publisher Services